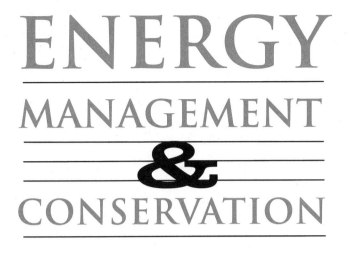

ENERGY
MANAGEMENT
&
CONSERVATION

FRANK KREITH

GEORGE BURMEISTER

National Conference of State Legislatures

National Conference of State Legislatures
1560 Broadway, Suite 700
Denver, Colorado 80202
William T. Pound, Executive Director

ISBN 1-55516-375-0

Energy Management & Conservation is printed on recycled paper.

CONTENTS

LIST OF TABLES AND FIGURES

FIGURE

LIST OF APPENDICES

FOREWORD

With the passage of the federal Energy Policy Act in October 1992, energy planning has risen in importance on state policymakers' agendas. The new legislation promotes long-range utility planning, economic development opportunities through energy efficiency, building code improvements, alternative fuels, and investments in renewable energy resources. States have been active in all of these areas for years, and this book documents their successful legislation and intiatives. In recent years some states have begun incorporating the social costs of energy into their planning decisions, and these efforts too are outlined in this important publication.

Energy Management & Conservation provides guidance and leadership to state legislators at a time when state revenue shortfalls and federal funding cuts to traditional state energy programs force us all to choose the best possible investments. As more states find it in their best interest to manage the efficient use of energy, we know that this book will serve as a blueprint for state action.

We are proud to acknowledge the support of the U.S. Department of Energy and the American Society of Mechanical Engineers in publishing this document, and we are pleased to present the results of this two-year collaborative research effort.

William T. Pound
Executive Director
National Conference of State Legislatures
December 1992

ACKNOWLEDGMENTS

The authors are grateful to the following reviewers who provided constructive comments and suggestions on the first draft of this document:

A.R. Adler, ASME Fellow for the State of New York; A.E. Bergles, 1990–1991 president of ASME; William Brennan, U.S. DOE; Clarence D. Council, WAPA; E.L. Daman, Foster Wheeler Development Corporation; Senator Patrick Deluhery (Iowa), 1991–92 chair, NCSL Environment Committee; Senator Vernon Ehlers (Michigan), 1990–91 chair, NCSL Environment Committee; Jonathan Koomey, Lawrence Berkeley Laboratory; Philip J. Mandel, U.S. DOE; Jim Miernyk, Western Interstate Energy Board; S.S. Penner, director of the Energy Center, UCSD; and Rebecca Vories, Infinite Energy, Inc.

Section VIII, which deals with the social costs of energy, draws extensively on Jonathan Koomey's work—particularly LBL Report No. 28313. However, interpretations and averaging of the data in his report are entirely the authors'.

R.E. West, of the University of Colorado and Nathan H. Hurt, Jr., 1991–92 president of ASME, proofread the final draft of part 1 in its entirety and offered many helpful suggestions that improved the final version. Janet Taylor of the Illinois Auditor General's Office reviewed the criteria for evaluating conservation measures and offered helpful comments. Beverly Weiler helped type the manuscript. Kristin Rahenkamp and Eileen Josslyn, ASME interns at NCSL, helped with tables, calculations, and references, and Eric Sikkema contributed a diligent summary of the 1992 Energy Policy Act and helped with editing. The American Society of Mechanical Engineers' State and Local Government Program supported Dr. Kreith's fellowship at NCSL while he worked on *Energy Management & Conservation*. Leann Stelzer not only typed most of the manuscript, she guided it through a number of rounds of editing and eliminated errors and inconsistencies. The project's success is due, in large part, to her vigilance.

Energy Management & Conservation was also made possible by the cooperation and support of many people within both the legislative and executive branches of state government, and by public utility commission staff. In genuine collaborative spirit, they devoted hundreds of hours to this endeavor. We wish to thank those who assisted with the writing of this book:

In Florida: B. Gene Baker, Division of Legislative Library Services; Wayne Voit, House Natural Resources Committee; Tom Batchelor, House Committee on Regulated Services and Technology; Ann Stanton, Department of Community Affairs; Richard Shine, Public Service Commission; and Kate

Nielson-Nunez , Executive Office of the Governor.

In Iowa: Kathy Hanlon, Legislative Services Division; Paul Erickson and Roya Stanley, Department of Natural Resources; and Gordon Dunn, Utilities Board.

In Nevada: Don Bayer, State Legislature; Curtis Framel, State Energy Conservation Program; Chris K. Freeman, Nevada Department of Conservation and Natural Resources; and Tom Henderson and Frank McRae, Public Service Commission.

In New York: the late Assemblyman Bill Hoyt; Ruth Horton, Peter Smith, and Bill Saxonis, Energy Office; James Gallagher, Public Service Commission; and Arnold R. Adler, Legislative Commission on Science and Technology.

In Washington: Frederick S. Adair, House of Representatives; Dick Byers and Michael A. Farley, Energy Office; and Deborah Ross, Utilities and Transportation Commission.

In Wisconsin: John Stolzenberg and David Lovell, Legislative Council; Don Wichert, Energy Bureau; Curtis Gear, Community Dynamics Institute; Ergun Somerson, Department of Industry, Labor and Human Relations; and Nancy Korda and Paul Newman, Public Service Commission.

Special thanks to Sigrid Higdon, U.S. Department of Energy, who provided important insight and guidance during the year this book was written; and to Frank Stewart, U.S. Department of Energy, for supporting NCSL. We are deeply indebted to Dr. Peter Blair, Office of Technology Assessment; to Clarence Council, WAPA; to Jim Wolf, Alliance to Save Energy; and to Dr. Peter DeLeon, University of Colorado, for their time-consuming reviews of portions of this book. Thanks also to Robbie Robinson, of the Center for Applied Research, for collecting integrated resource planning details from various state public utility commissions.

Julie Poteet Wynn (Denver, Colo.) designed and illustrated the cover and the interior graphics and layout.

INTRODUCTION

"The future belongs to the efficient."

Admiral James D. Watkins, Secretary of Energy

This book gives an overview of energy-management and conservation efforts by state governments. Its objective is to help state legislators implement effective energy management on a state or regional level that will reduce energy consumption by improving energy-use efficiency.

The book is divided into two parts. The first is general in nature. It discusses current and projected United States energy use, summarizes the National Energy Strategy (NES), presents the results of a National Conference of State Legislatures (NCSL) survey designed to rank state energy issues, reviews attempts to assess social cost of energy use quantitatively, and discusses integrated resource planning (IRP) efforts by the states. It presents quantitative and qualitative criteria for evaluating energy-conservation and efficiency programs undertaken by the states and gives a synopsis of future options and opportunities for energy savings by state legislation. The last chapter of part one provides a long-term perspective of the energy problem.

To prepare an effective energy-management plan on a state level and implement IRP that considers energy production, load shaping, and demand-side reduction on the same basis, one needs to know the equivalent costs of energy-conservation and efficiency measures in cents per kilowatt-hour (c/kWh), as well as the qualitative aspect of implementing demand-side reduction. Part two of this work presents an evaluation of the effectiveness of energy-efficiency and conservation programs undertaken by six states during the past 25 years; that is, since the first energy crisis in the 1970s. These program evaluations can provide a data base for energy decisionmakers by determining what types of programs have been effective (including their cost and payback time), and what types have not achieved energy savings in a cost-effective manner. The organization of part two follows the same pattern for all six states. The first section introduces the specific situation of the state, its past energy use, and future projection; the second section discusses energy use by sectors—residential, commercial, industrial, and transportation; the third section gives a synopsis of the energy legislation in the state; the fourth describes the state's energy efficiency programs or initiatives, some of which are evaluated in a matrix at the end of section four. The matrix uses qualitative and quantitative criteria to evaluate the energy programs. The last chapter presents an evaluation summary of the programs in the six states, as well as some conclusions and recommendations.

UNITED STATES
ENERGY USE EFFICIENCY

"Just as there must be a balance in what a
community produces, there must be a balance in
what the community consumes."

John Kenneth Galbraith

E nergy is a mainstay of any industrial society. It is an essential input
to the economy to run the country's factories and provide such
comforts as mobility, heat, and light. Using energy also generates
pollution and health problems. Inefficient use depletes limited natural
resources unnecessarily, hurts our balance of trade, and reduces the national
security.

Until recently, energy policy was considered largely the responsibility
of the federal government. During the past 10 years, however, primary
responsibility for energy regulation and planning has shifted from the federal
government to individual states and there is a growing realization at the state
level that energy is a scarce commodity that must be used efficiently, and that
state laws and state agency programs can be used to achieve this goal.

Figure 1 shows that from 1950 to 1971 energy use and gross domestic
product (GDP) in the United States increased at roughly the same average
annual rate (3.5 percent).[1] As a result, it was widely assumed that economic
growth was linked to increased energy consumption. But this apparent link
unraveled between 1972 and 1985 when the average growth rate of GDP was
2.5 percent while energy use increased at an annual rate of only 0.3 percent. In
that period, 20 million homes were added to the country's housing stock, the
fleet of vehicles on America's roadways increased by 50 million, the number of
business establishments rose by 1.5 million, the GDP grew by 39 percent in
real terms, and the amount of energy used to produce one dollar's worth of
economic output fell by 2.4 percent per year. Unfortunately, however, the

improvement in energy efficiency in the United States ceased in 1985. Between 1985 and 1988, an 8 percent per annum growth in energy consumption occurred, while the energy intensity (Btu/GDP) barely changed.

Figure 1

Index of United States Energy Use, GDP and Energy Intensity

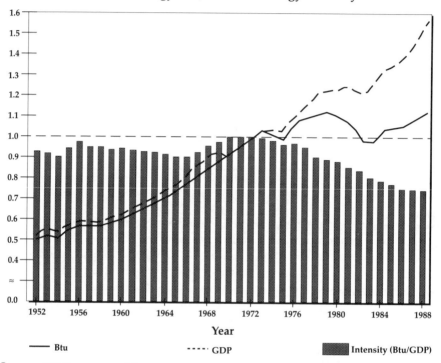

Source: U.S. Department of Commerce, Bureau of Economic Analysis, National Income and Product Accounts, table 1.2; U.S. Department of Energy, Energy Information Agency, *Annual Energy Review 1987*, table 4; and the *Monthly Energy Review*, August 1989, table 1.4.

The Office of Technology Assessment (OTA) attributed these changes in energy intensity over time to a combination of increases in energy-use efficiency and structural changes in the United States economy.[2] The recent growth of the service sector, as well as implementation of energy-conservation measures at the state level, have contributed to the decline in U.S. energy intensity. Also, the leveling out of the cost of oil, after the oil crises of 1973 and 1983, and the overall slowdown in the economy contributed to the decline. Reasons for the turnabout from a steady decrease in the energy-use-to-GNP ratio to a sudden increase in 1985 are the subject of debate. The economy had recovered from the previous recession and indications are that about that time many of the most cost-effective energy-conservation efforts had been put in place, leading to a decline in investment for energy-conservation measures by industry and consumers.

Future U.S. energy trends are difficult to predict. They will depend largely on relative growths in the service and industrial sectors, as well as on the successful implementation of energy-conservation measures at the state and national levels. These issues are considered further in chapter 4. To compete on the world market, there is a pressing need to improve energy-use efficiency and productivity in the United States. To produce one unit of GNP, Western Europe consumes 57 percent of the energy used by the United States to produce the same unit; Japan consumes 44 percent. Because energy cost is an important ingredient in the price of goods, improved energy-use efficiency will make U.S. products more competitive.

Measures taken at the state level are often credited for a significant portion of the success in reduced energy consumption. State governments spent hundreds of millions of dollars on energy-efficiency programs over the last decade. Funding for these was provided by state governments, the federal government, and private foundations. However, the majority of funding came from oil-overcharge funds, distributed by the U.S. Department of Energy (DOE) over the last five years. These funds—which reached more than $4 billion—resulted from a 1988 settlement that resolved alleged violations of the petroleum price and allocation controls in effect between August 1973 and January 1981.[3] Subsequent legislation by U.S. Senator John Warner (Va.) in 1982 (Public Law [PL] 97-372) provided for a one-time allocation of oil-overcharge money to the states, proportional to each state's oil use during the period the alleged overcharge took place. Warner's bill specified that the oil-overcharge money given to the states must be appropriated among five previously existing federally funded state energy grant programs: the State Energy Conservation Program (SECP), the Weatherization Assistance Program (WAP), the Energy Extension Service (EES), the Institutional Conservation Program (ICP), and the Low-Income Home Energy Assistance Program (LIHEAP). Warner's bill also specified that the general public receive restitution through energy savings achieved by these programs.

The Safe Energy Council estimates that energy-efficiency measures implemented at the state level since 1972 currently save the United States more than $170 billion annually.[4] With state budget constraints increasing and the absence of a national energy strategy that mandates conservation and improvements in energy-use efficiency, state legislatures have initiated their own legislation to reduce energy use.[5] This book describes the status of state efforts to increase energy efficiency, to reduce oil consumption, and to encourage conservation at state and local levels. It also indicates future opportunities to further these goals.

6

Chapter Notes

U.S. Department of Commerce, *National Income and Products Accounts* (Washington, D.C.: 1989), table 1.2.

L.S. Johns and P.D. Blair, *Energy Use and the U.S. Economy* (OTA-BP-57, Washington, D.C.: U.S. GPO, June 1990).

T.L. Johansen, "An Analysis of Wisconsin's Oil Overcharge Funded Energy Programs," M.S. thesis (University of Wisconsin: Madison, 1990).

Safe Energy Communication Council, *SECC's Mythbusters No. 6* Fall 1990: 2.

K.D. Sibold, et al., *Iowa at the Crossroads: 1990 Iowa Comprehensive Energy Plan* (Iowa Department of Natural Resources, January 1990).

THE NATIONAL ENERGY STRATEGY

> "By 1980 we will be self-sufficient and will
> not need to rely on foreign energy."
>
> *Richard M. Nixon*
> *proclaiming "Project Independence" in 1973*

The knowledge that a secure and adequate supply of energy is essential for the economic health of the United States, coupled to recognition that indigenous sources of liquid fuel are declining and an increasing portion of our earnings goes to pay for oil imports from a highly volatile part of the world, has led to several attempts to establish a national energy policy. Following the Arab oil embargo in 1973, President Nixon initiated energy planning in an effort to make the country energy independent. In 1977, the 95th Congress mandated the periodic preparation of energy plans (DOE Energy Organization Act, PL 95-91). In 1977, President Carter responded with the National Energy Plan, updated to the National Energy Plan II in 1979. This plan attempted to achieve energy independence and develop domestic energy sources, including oil shale, wind, biomass, and solar energy. President Reagan submitted National Energy Policy Plans in 1981, 1983, and 1985, followed by the Energy Security Plan in 1987. These were based on the assumption that a free market would provide a secure and adequate supply of energy for the nation.

The latest efforts to develop a National Energy Strategy (NES) began in the summer of 1989, when five national laboratories were convened by the DOE to discuss analytical requirements associated with the development of an NES. Beginning August 1, 1989, a series of 16 public hearings were initiated, by which DOE solicited information and opinions from the American public about problems, solutions, and priorities associated with future energy supply and demand.

Between January and March 1990, DOE national laboratories released a series of white paper reports that provided technical and analytical background for the NES; these addressed five topical areas: future trends in end-use energy consumption,[1] the potential for renewable energy,[2] energy and climate change,[3] energy technology for developing countries,[4] and technology transfer.[5] In April 1990—after the last public hearing on the NES was held in Washington, D.C.—DOE released its Interim Report on the National Energy Strategy.[6] The Interim Report reflects the comments of 379 public hearing participants and over 1,000 written comments.

President Bush released the 1991 version of the NES in February 1991.[7] It is the culmination of more than 18 months of public hearings and debates to achieve greater efficiency in the use of energy and spur competition in energy markets while balancing environmental and economic interests. Secretary of Energy Admiral James D. Watkins said, "The NES is a strategy that, for the first time in our nation's history, lays a comprehensive foundation for a cleaner, more efficient and more secure energy future. It is a balanced approach to attaining that energy future through shared responsibility."[8] The highlights of DOE's proposed NES are summarized in **table 1**.[9] Various reactions to the Interim Report and the 1991 version of the NES are given in references 10–12.

Table 1
Highlights of DOE's National Energy Strategy Proposed in 1991

Energy conservation and efficiency
- Require that the energy secretary publish energy conservation standards for electric lights; require the Federal Trade Commission to prescribe rules for labeling lights.
- Authorize loans from the U.S. Treasury for energy conservation measures at government agencies.
- Remove taxes from rebates awarded by utilities to customers who install high-efficiency lighting and appliances.

Nuclear power
- Require the Nuclear Regulatory Commission (NRC) to consolidate construction and operating licensing into a single procedure incorporating inspection, testing and analysis.
- Require the NRC to resolve emergency planning before issuing the combined license; specifically, to disallow after issuance of the license any hearings based on "a decision of a state or local government to withdraw from participation in emergency planning."

Nuclear waste
- Allow the energy secretary to conduct, without a state or local permit, "site characterization studies" under the Nuclear Waste Policy Act of 1982, but also require the secretary to "consider the views of state and local officials" with regard to local laws affecting site characterization.

Renewable energy
- Provide tax credits of up to 2 c/kWh for those producing electricity with solar-thermal, photovoltaics, wind or "certain geothermal" and biomass technologies.
- Remove power production limitations (80 MW) from eligible alternative-power plants.

Table 1 (continued)

Highlights of DOE's Proposed National Energy Strategy

Reform of Public Utility Holding Company Act
• Allow both utilities and non-utility organizations to use a holding company structure to build and finance independent power projects not subject to the financial and corporate regulations of the Holding Company Act.

Transportation
• Require that a fraction of all automobile and truck fleets assembled after 1994 run on alternative (non-petroleum-based) fuels.

Hydroelectric power regulatory reform
• Require the Federal Energy Regulatory Commission (FERC) to coordinate and consolidate licensing review for hydroelectric plants, taking into account the inputs of federal and state agencies and Native American tribes.
• Exempt projects that produce less than 5 MW from licensing requirements.

Energy supplies
• Establish a program to lease lands in the Arctic National Wildlife Refuge's 600,000 hectare coastal plain for oil and gas exploration and drilling.

Fuel regulation
• Replace existing FERC regulatory authority over oil pipelines with "streamlined common carrier obligations."
• Allow import and export of natural gas without prior Federal government approval.
• Give the FERC sole jurisdiction over environmental impact statements for gas pipelines.

Source: IEEE Spectrum, March 1991.

Controversial provisions in the 1991 NES relating to the Arctic National Wildlife Refuge (ANWR) and corporate average fuel economy (CAFE) standards halted congressional action on the NES in 1991. However, the NES and its original concepts were redrafted in 1992 and included as part of the National Energy Security Act of 1992, which was subsequently passed by the U.S. Congress in October 1992 as the Energy Policy Act of 1992. Highlights of this legislation are included in appendix 11.

Chapter Notes

1   U.S. Department of Energy, *The Interim Report on the National Energy Strategy* (Washington, D.C.: April 1990).

2   U.S. Department of Energy, *The National Energy Strategy* (Washington, D.C.: February 1991).

3   M. McCracken, *Energy and Climate Change*, U.S. Department of Energy White Paper (Livermore: Lawrence Livermore National Laboratory, January 1990).

4   D. Doll, *Energy Technology for Developing Countries*, U.S. Department of Energy White Paper (Livermore: Lawrence Livermore National Laboratory, January 1990).

5   D. Deonigi, *The Technology Transfer Process*, U.S. Department of Energy White Paper (Richland, Wash.: Pacific Northwest Laboratory, January 1990).

6   *The Interim Report on the National Energy Strategy*, April 1990.

7   *The National Energy Strategy*, February 1991.

8   *Colorado Energy Talk*, May/June 1991.

9.   IEEE *Spectrum*, March 1991.

10.   Western Interstate Energy Board, *Western Energy Update 91*, 5 (Denver: Western Interstate Energy Board, March 15, 1991).

11.   A.E. Bergles, "A Closer Look at the National Energy Strategy," *ASME News 10*, April 2, 1991.

12.   P.H. Abelson, "National Energy Strategy," *Science* 251, 5000 (March 22, 1991): 1405.

PROJECTIONS FOR UNITED STATES ENERGY CONSUMPTION

"Alaska...has a greater oil reserve than Saudi Arabia."

Ronald Reagan, quoted in the Washington Post, February 20, 1980.

In June 1990, the U.S. General Accounting Office (GAO) prepared a report for congressional committees involved with energy and environmental issues.[1] It summarized numerous recent energy-related publications by GAO and was intended as background for DOE development of its NES. Previous projections for future energy use, particularly for electrical power, have not always proved valid, but a database is needed on which to develop an energy policy for the coming decade. Although the energy picture varies from state to state, the overall perspective of the GAO may be useful to state legislators.

It is important to remember that in recent years energy issues have become closely linked to environmental concerns caused by the use of fossil fuels that contribute to acid rain, air and water pollution, and global warming. Although coal reserves in the United States are estimated to last several centuries, the rate at which coal can be used as a fuel in the future will depend to a large measure on successful reduction of environmentally damaging emissions from coal-fired generators, especially sulfur dioxide (SO_2) and nitrogen oxide (NOX), which contribute to acid rain. In addition, concern about carbon dioxide (CO_2) emission, which causes global warming and could have serious adverse environmental effects, may limit the rate at which coal can be used. Nuclear power technology is available to meet future base-load electric power needs, but the public's concerns about its health and safety may impede its increased use. Also, there exists no safe repository for long-term storage of spent nuclear fuel and the public has deep-rooted fears about the disposal of high-level radioactive wastes.[2] These and other concerns demand that state policymakers consider environmental impacts as well as cost and availability when deciding future energy-choice policies.

Figure 2 shows U.S. consumption of energy by source since 1973, as well as Energy Information Agency (EIA) projections through 2000. Since 1973, the nation's consumption of nuclear power, coal, and oil increased, while natural gas consumption declined. The United States is expected to make use of a wide range of available energy resources in the future, but the percentage that each source will contribute to total energy needs will undergo important changes. According to EIA, natural gas consumption is expected to increase substantially, contributing about 25 percent of total U.S. energy use by 2000. There has been some concern about future supply,[3] but according to estimates by DOE and the Department of Interior, natural gas reserves in the United States are expected to last until the middle of the next century at current rates of consumption,[4] thus giving policymakers confidence about the availability of this fuel. Natural gas combustion is relatively clean of soot, carbon monoxide (CO), sulfur oxides (SO), and NOX generated by burning coal and oil. The emission of CO_2, which is the most important of the greenhouse

Figure 2

Total United States Energy Consumption by Source

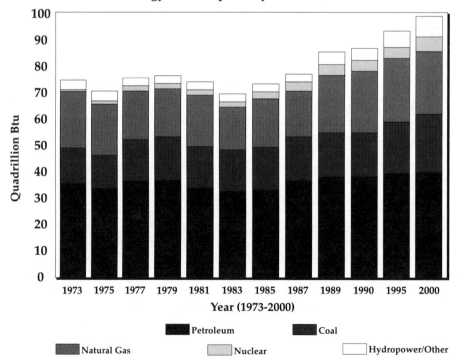

Source: Cited in *Energy Policy: Developing Strategies for Energy Policies in the 1990s* (GAO/RCED-90-45), June 1990.

gases that cause global warming, is much less from natural gas combustion than from coal or from oil per unit of energy output.[5] But natural gas is not only in finite supply, it is an important feedstock for the chemical industry. Hence, there is concern that it is too valuable to be burned for power that could be obtained from other sources.[6]

Renewable sources such as hydropower, biomass, geothermal, wind, and solar have great popular appeal. EIA expects consumption of hydroelectric power to increase appreciably. Hydroelectric plants are efficient, cause no air pollution in operation, and deliver electricity at relatively low cost. **Figure 3** shows the resource availability of hydroelectric power in the United States by regions, according to a 1988 study by FERC.[7] The largest resource potential is in the Pacific region (22.1 million kW), followed by the mountain region (20 million kW). There are currently 2,029 developed hydroelectric plants in the United States with a hydroelectric capacity of over 40,000 MW and a pumped storage capacity of 17,000 MW,

Figure 3

Developed and Undeveloped Hydroelectric Resources by Geographic Region

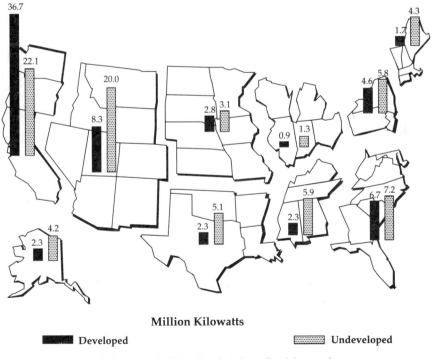

Million Kilowatts

■ Developed   ▦ Undeveloped

Note: The hydro potential of Hawaii is relatively small and thus not shown.

Source: Federal Energy Regulatory Commission, 1988.

capable of supplying about three quads* per year. Some existing hydroelectric facilities can generate power at less than 1 c/kWh, but the cost from new hydro capacity is estimated between 4 c/kWh and 9 c/kWh in levelized current dollars. Power from small hydroelectric plants (with capacity of less than 30 MW) is more expensive than from large plants, but there are many more potential sites for smaller plants. There are, however, obstacles to new hydroelectric developments; J. Miernyk of the Western Interstate Energy Board (WIEB) believes that "large hydroelectric plants will experience great difficulty in obtaining permits due to loss of land area and destruction of natural habitat."[8]

Municipal solid waste (MSW), which contains biomass in the form of food, wood, and yard waste, is expected to be used as an energy source in more waste-to-energy conversion plants. There are currently more than 100 such plants in operation worldwide. The decrease in available landfill capacity, coupled with the opposition of the public to building new landfills (the NIMBY [not in my backyard] syndrome), is likely to force municipal and state governments to use incineration to reduce the volume of the waste. Recycling is the least controversial means to dispose of waste; but even if a 25-percent recycling rate were achieved, incineration of waste would still be required to reduce the volume of the remaining 150 million tons per annum MSW predicted for the year 2000. This waste could supply between 2 percent and 3 percent of the total electric power needed in the United States.[9] Florida has legislation that promotes burning waste over disposing it in a landfill (see part II, Florida).

Use of wind power, which already supplies more than 1 percent of California's electricity, may also increase in favorable locations. According to R. Swisher, 16 states have wind energy potential equal to, or greater than, California. These states, in order of greatest potential for wind power, are North Dakota, Texas, Kansas, Montana, South Dakota, Nebraska, Wyoming, Oklahoma, Minnesota, Iowa, Colorado, New Mexico, Idaho, New York, Illinois, and Michigan.[10] Capital costs for commercial wind systems are now as low as $1,000 per kW of capacity and costs for operation and maintenance are about 1 c/kWh, according to Swisher. Electric wind power costs are between 6 c/kWh and 9 c/kWh, depending on the wind speed at the site. But solar and other renewable resources are not expected to provide substantial amounts of power by the end of the century unless the federal government provides financial support or tax subsidies. States with good potential for solar energy are California, Arizona, Florida, Nevada, Texas, New Mexico, and Colorado.

---

*One quad of heat equals about 170 million barrels of oil, one trillion cubic feet of natural gas or about 45 million short tons of coal. In 1989, United States domestic energy consumption was about 84 quads.

Electric power is important for an industrial nation. For the past 30 years, electricity use in the United States has grown at a faster rate than the economy. Data complied by the departments of Energy and Commerce show that between 1973 and 1985, the ratio between the growth rates of electricity use and GNP was approximately 1 to 1; from 1986 through 1990, electricity use grew 3.6 percent annually, while the GNP rose 2.8 percent per year.[11]

During the same period (1973-1990), the overall ratio of energy consumption to GNP ratio declined by almost 30 percent. Some economists attribute the rise in the ratio between electricity use and overall energy consumption to improved industrial productivity.[12] Others claim that it is due to aggressive advertisement by utilities for new uses, such as air conditioning, and the convenience of using electricity in the building sector. The public often perceives electricity use as environmentally more benign than oil or coal burning because the environmental effects are not visible at the point of consumption.

According to the U.S. Council for Energy Awareness, electricity use has grown faster than the GNP since 1960 as a result of the continuing "electrification" of the United States economy—that is, use of electricity instead of direct burning of fossil fuels like oil, natural gas, and coal.[13] In 1990, electricity accounted for 36 percent of all the primary energy used in the United States, up from 19 percent in 1960. The EIA predicts that by 2010 electricity will account for 40 percent of the nation's total primary energy consumption. Some analysts attribute the increasing electrification of the United States economy in part to the 24 percent decline since 1984 in "real" electricity prices (that is, discounted for inflation) and the rebound since the mid-1980s of electricity-intensive industries, such as steel, aluminum, and chemicals. For example, small steel mills, which use electric arc furnaces, now account for 36 percent to 37 percent of all U.S. steel production, and they may account for up to 40 percent by the turn of the century.

The EIA's 1991 *Annual Energy Outlook* assumes that annual growth in electricity sales and GNP will be 2 percent and 2.1 percent, respectively, through the year 2010. The NES, however, assumes that electricity sales will grow *significantly* slower than the economy over the next 20 years.[14] From 1990 to 2000, the NES projects annual growth of 2.6 percent in electricity use and 3.2 percent in GNP. From 2000 to 2010, the NES projects annual growth of 1.4 percent in electricity sales and 2.8 percent in GNP. According to this scenario, the NES projects that from 190,000 to 275,000 MW of new electric generating capacity will be needed by 2010. The Nuclear Power Oversight Committee expects that the demand for electricity will require at least 100,000 MW of new capacity by the end of the century.[15] However, there are indications that the annual growth rate of 3.2 percent in GNP is overly optimistic and that the slower rate predicted by EIA for electricity sales is more realistic. Expanded demand-side reduction of electric power could affect the above predictions and reduce the need for new generating capacity.

The most serious problem concerns the future supply of oil. Domestic crude oil production increased from 1979 to 1985 as a result of high world oil prices, but the decline in oil prices in 1986 reversed this trend. Lower oil prices discouraged investment in oil exploration and development. It also led to the shutdown of cost-intensive stripper wells that had accounted for nearly 15 percent of domestic production. Since 1985, domestic crude oil production has decreased by about one million barrels per day and the EIA expects this trend to continue. In 1989, domestic oil production was approximately 7.7 million barrels per day and is expected to decline to 5.9 million barrels per day by 2000. To make up this shortfall and anticipated growth, EIA expects net daily imports to increase from 7.2 million barrels per day to almost 10 million barrels per day by 2000. According to DOE, the United States has approximately 300 billion barrels of oil remaining in the ground. Some of this oil, however, will require the use of relatively expensive secondary or tertiary oil-recovery techniques, while other oil is located in smaller, remote fields that are costly to explore and develop.

Figure 4 shows EIA projections that indicate the nation's trend toward increased oil consumption will continue largely because of increased consumption by the transportation sector. The transportation industries sector, which consumed nearly 11 million barrels of oil per day in 1989—two-thirds of total U.S. oil consumption—is wholly dependent on petroleum fuels and lacks ability to switch rapidly to any other fuel. According to GAO, policymakers will have to emphasize energy efficiency in future energy policies, particularly in the transportation sector.

Efforts are underway in the U.S. Congress to save gas and oil by improving the mileage of cars in the future. For example, the Motor Vehicle Efficiency Act of 1991, sponsored by Senator Richard Bryan (Nev.), was designed to save oil nationwide by raising the CAFE standards from today's level of 27.5 fleet mpg to 40 fleet mpg by the year 2000. According to Bryan, the United States would save 2.5 million barrels of oil per day once the new CAFE standards were fully implemented. The bill also contained a "truth-in-testing" amendment to ensure accuracy in the government mpg rating, because current tests are said to exaggerate fuel efficiency by 15 percent.

The Bryan bill was defeated by three votes in the Senate. Opponents of the Bryan bill cited the poor safety record of smaller, fuel-efficient cars and a lack of demand from the American public as reasons to keep the CAFE constant, if not roll it back. Chrysler lobbyist Albert Slector contends that if Americans were forced to buy smaller cars it would lead to increased traffic deaths. However, the Sierra Club and some consumer advocates claim that the domestic auto manufacturers have failed to exploit modern technology. They cite *The Safe Road to Fuel Economy*,[16] a report issued in April 1991 by the Center for Auto Safety and MCR Technology Inc., which concluded that, "with utilization of the advanced safety and fuel economy

Figure 4
Oil Consumption by End-use

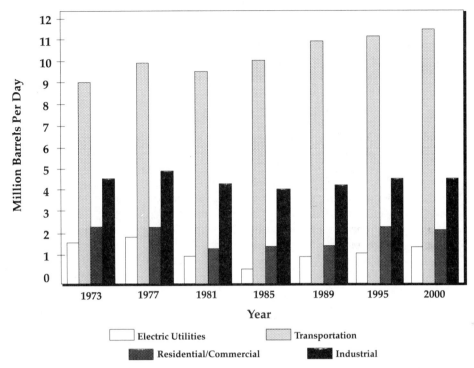

Source: Cited in *Energy Policy: Developing Strategies for Energy Policies in the 1990's* (GAO/RCED-90-45), June 1990.

technologies (discussed in the report), CAFE levels of 40-50 mpg can be attained by 2001 without any shift to smaller cars." Although state governments have little influence on CAFE standards, state legislation such as a revenue-neutral "gas guzzler tax," that would create economic incentives to buy fuel-efficient cars, could achieve the same goal.[17] State legislators are interested in incentives like those contained in Maryland's new (1992) "fee-bate" legislation.

Alternative Fuels

The development of alternative fuels to gasoline, such as methanol and ethanol, are other options to be considered. However, these alternative fuels are likely to be more expensive than current gasoline. Estimates for production of ethanol from corn are about $66 per barrel of oil equivalent (bble) and for methanol from coal about $52 per bble.[18]

Ethanol. Ethanol is used as a motor vehicle fuel in some other countries, particularly Brazil. In the United States, ethanol is produced primarily by fermentation of sugar which, in turn, is obtained by hydrolysis of starch from grain (mainly corn). The by-product of ethanol production is a mixture of fibers and proteins that can be used for animal feed. Ethanol is used in this country primarily as a blend in gasoline at concentrations of 7 percent to 10 percent, often called gasohol. Even when credit is given for the octane-enhancing properties of ethanol, its economic competitiveness is secured only by tax subsidies. Moreover, substantial amounts of oil are required to produce the fertilizers used to grow corn. Because of the uncertainty of the future of these subsidies, the U.S. capacity for production of fuel grade ethanol has grown slowly, and is significantly less than 1 percent of the gasoline consumed nationwide.

Methanol. Methanol is currently produced mainly from natural gas, but it can also be produced from coal using established technological processes. Methanol can also be produced from well-head gas, an often undesirable by-product of pumping crude oil. The process is similar to the technology used for producing methanol from natural gas and would negate the necessity of flaring well-head gas—a practice not prevalent in the United States, but common elsewhere. Methanol produced from processing of well-head gas offers a potential source of liquid fuel.

In contrast to ethanol, fuel methanol can be used at concentrations ranging from 85 percent to 100 percent, often referred to as neat methanol, and provides significant environmental advantages. Considerable experimentation has been carried out with neat methanol as a fuel. Potential environmental benefits include decreases in the emission of oxides of nitrogen, carbon monoxide, and particulate matter, as well as the replacement of the unburned hydrocarbons produced by gasoline with methanol, which is relatively unreactive in the formation of photochemical smog. Further, a slight increase in fuel efficiency is achieved.

The three major U.S. car manufacturers have supported methanol as the most logical replacement for gasoline and important programs are being pursued for the development of methanol vehicles. But current production capacity for methanol is too small to meet a significant share of the demand, should neat methanol be used as a substitute for gasoline.

Compressed natural gas. The alternative fuel most likely to be used on a large scale in the near term is compressed natural gas (CNG). Vehicles that store natural gas in tanks at high pressure are the most common type of alternative-fuel vehicles on the road today. Half a million CNG vehicles are in operation worldwide—300,000 in Italy, 120,000 in New Zealand, 30,000 in the United States, and 17,000 in Canada. Most CNG is used in buses, delivery vehicles and government fleets. Natural gas companies use half of such vehicles in the United States today.

CNG, like all alternative fuels, has various advantages and disadvantages. While existing CNG-fueled vehicles are relatively more expensive than gasoline or methanol vehicles, this cost could be offset by lower fuel and maintenance costs. CNG-fueled vehicles are also relatively safer than gasoline vehicles in cases of tank rupture. However, current CNG vehicles have a limited driving range of approximately 200 miles and there are only a few places where they can be filled with fuel. Consequently, some CNG vehicles are "dual fueled," meaning that they also contain a gasoline tank and can switch back and forth between the two fuels, which makes them more convenient to drive.

The Alternative Motor Fuels Act of 1988

Congress' intent for the Alternative Motor Fuels Act (AMFA) was to improve the competitive position of alternative transportation fuels compared to conventional fuels. AMFA is intended specifically to promote the use of methanol, ethanol, and natural gas, but does not mandate a program that will put a specific number of alternative fuel vehicles (AFV) on the road in a given time period. It assigns to DOE— with the assistance of GSA, DOT, and EPA— the task of encouraging adoption of AFVs. However, the success of the AMFA program hinges on the participation of states and local governments, as well as federal agencies. It focuses on demonstrations and tests of alternative fuels in three categories—federal light-duty vehicles, heavy-duty trucks, and buses. In the first category, Ford Motor Company and General Motors Corporation have provided 65 methanol flexible-fuel vehicles to the federal fleet. The heavy-duty truck project provides joint funding demonstration projects with state governments and the private sector to augment existing state and private heavy-duty fleet tests. The bus project helps state and local governments test alternative-fuel buses in cities. This project may ultimately involve more than 50 state and local transit agencies throughout the United States.

Electric Vehicles

Electric vehicles have been proposed as an alternative to the current internal combustion automobile. The key drawbacks of electric vehicles are their high cost, limited range, the time required to recharge batteries, and the expense of periodic replacement of battery packs. However, regulations in pollution-conscious states like California may provide sufficient incentives for automobile makers to enter the electric vehicle market. The California Air Resources Board recently mandated that any auto maker who plans to sell 5,000 or more cars per year in the state must sell zero-emission vehicles (ZEV) by 1998. The mandate initially requires that 2 percent of the manufacturer's

California sales be ZEVs; but by 2001, ZEVs must account for 5 percent of sales and increase to 10 percent by 2003.

With this huge market at stake (about 100,000 cars), U.S. auto makers, as well as European and Japanese companies, have begun to develop electric vehicles. Another near-term target comes from the Los Angeles Electric Vehicles Initiative, aimed at putting 10,000 electric vehicles on roads by 1995. Other states—such as New York, New Jersey, Pennsylvania, and Massachusetts—recently announced plans to adopt the tough new California clean air rules and may also provide markets for electric vehicles.

Among the automobile manufacturers entering the electric vehicle market are VW, with a hybrid vehicle called the Chico; BMW, which recently converted a 325iX sedan to electric propulsion; the Clean Air Transport AB, a Swedish-based company with engineering assistance from International Automotive Design of the U.K., which has developed the LA 301, a hybrid vehicle; and Nissan Motor Company, which claims to have developed the super-quick-charging batteries system that can recharge batteries in minutes instead of hours. Ford Motor Company, General Motors Corporation, and Chrysler Corporation have joined forces as the U.S. Advanced Battery Consortium to conduct research designed to improve the characteristics of electric vehicle battery technology.

An electric vehicle, however, still costs almost twice as much as an equivalent gasoline-powered car. For example, the two-seat GM Impact is estimated to cost between $15,000 and $20,000; battery packs, which must be replaced every 20,000 to 30,000 miles, may cost up to $8,000. Much development work will be needed before electric vehicles are competitive with other means of transportation; and stimuli, such as tax incentives, subsidies, or mandates, may be necessary before electric vehicles become a major force in the transportation field. Nevertheless, electric vehicles for short-range urban transportation are an option for the future, especially if base-load capacity from increased nuclear power production becomes available to charge car batteries. But if coal fired power plants are used to produce the electricity to charge the batteries, the air pollution produced will be substantial.

Energy use patterns and energy projections for individual states depend on geographic location, dominant industries, population density, and availability of public transportation. **Figure 5** shows the yearly per capita energy consumption in the eight states with the largest gross state products.[19] It can be seen that energy consumption varies widely, with New York using only 200 MBtu per capita in 1988, compared to 569 MBtu per capita in Texas. These large differences in *per-capita* energy consumption for different states reflect use patterns associated with the availability of public transportation (New York) and of low-cost indigenous gasoline supplies (Texas). Further, the dominance of service industries (New York) over heavy industrialization

Figure 5
Per Capita Energy Consumption for Select States, 1988

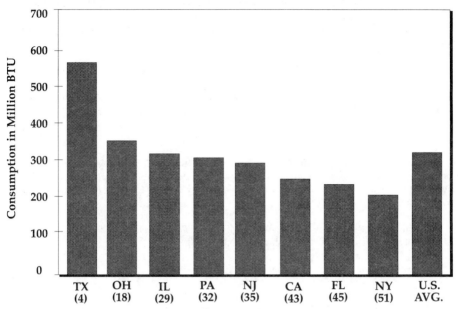

Note: State ranking in ( )
* The eight states selected are those with the largest gross state products.

Source: New York State Energy Office, "Energy Policy in the Aftermath of Kuwait," *Empire State Report*, April 1991, 21.

(Texas) greatly reduces *per-capita* energy use. Effective energy plans are best developed on an individual state basis to address local conservation issues and opportunities.

Chapter Notes

1   General Accounting Office, *Energy Policy: Developing Strategies for Energy Policies in the 1990s*, GAO Report to Congressional Committees (GAO/RCED-90–85), (Washington, D.C.: June 1990).

2   P. Slovik et al., "Perceived Risk, Trust, and the Politics of Nuclear Waste," *Science* 254 (1991): 1603–1607.

3   R. Kerr, "Natural Gas Estimates Plummet," *Science* 245 (September 1989): 1330–1331.

4   D.A. Dreyfus and A.B. Ashby, "Global Natural Gas Resources," *Energy –The International Journal* 14, 84 (1989): 773–784.

5   L. Morandi, *Global Climate Change* (Denver: NCSL, 1990).

6   A.E. Bergles, "Statement of the Ad Hoc Task Force of the ASME Energy Resources and Conversion Groups on Meeting Long-Term Energy Policy Objectives to the Subcommittee on Energy of the Committee on Science, Space and Technology of the U.S. House of Representatives," (Washington, D.C.: June 16, 1991).

7   Federal Energy Regulatory Commission, *Hydroelectric Power Resources of the U.S., Developed and Undeveloped* (Doc. FERC-0070) (Washington, D.C.: 1988).

8   J. Miernyk, October 7, 1991: personal communication.

9   F. Kreith, "Solid Waste Management in the U.S. and 1989–91 State Legislation," *Energy –The International Journal.*, 17 (1992): 427–476. See also S.S. Penner and M.B. Richards, "Estimates of Growth Rates for Municipal Waste Incineration and Eventual Control Costs for Coal Utilization in the U.S.," *Energy –The International Journal* 14 (1989): 961–963.

10   R. Swisher, "Wind Energy Comes of Age,"*Solar Today* (May/June 1991): 14–17.

11   "U.S. Electricity Sales Still Rising Faster than the Economy," INFO, U.S. Council for Energy Awareness, 267 (August 1991).

12   E.L. Daman, "Energy Policy for the 21st Century," (paper delivered at 1991 ASME Pressure Vessels and Piping Conference, San Diego, June 1991).

13   "U.S. Electricity Sales Still Rising Faster than the Economy."

14   U.S. Department of Energy, *The National Energy Strategy* (Washington, D.C.: February 1991).

15   Nuclear Power Oversight Committee, *A Perfect Match: Nuclear Energy and the National Energy Strategy* (Washington, D.C.: November 1990).

16   Center for Auto Safety and MCR Technology, Inc., *The Safe Road to Fuel Economy* (April 1991). See also *Colorado Daily* editorial, May 13, 1991.

17   J. Koomey and A.H. Rosenfeld, "Revenue-Neutral Incentives for Efficiency and Environmental Quality," *Contemporary Policy Issues* 8 (July 1990): 142–155.

18 *Fuels to Drive Our Future* (Washington, D.C.: The National Academy Press, 1990).

19 New York State Energy Office (NYSEO), "Energy Policy in the Aftermath of Kuwait," *Empire State Report* (Albany: April 1991): 21.

NCSL STATE SURVEY OF ENERGY ISSUES

"All men by nature desire knowledge."

Aristotle

In an effort to gauge the mood and direction of the country, NCSL each year surveys legislatures in the 50 states, in Puerto Rico, and in the District of Columbia. The goal of the survey, which covers nearly 1,000 issues in 18 topic areas, is to identify priority issues for the coming legislative session. More than 1,400 survey responses from several hundred state legislators and legislative staff are then summarized in the publication, *State Issues*. The broad area of energy issues is subdivided into two sections in the survey: energy issues and electric utility issues. This section presents a combined summary of the results of these two areas for 1991. Results from the 1992 survey by George Burmeister indicate increased interest in energy efficiency and decreased interest in production-oriented issues.[1]

In the energy area, the NCSL survey indicated that development of safety regulations, particularly for above-ground storage tanks, oil and gas development, and oil tankers, were considered high priority. This reflects growing concern for environmental and human safety with respect to energy development; at least 18 states expect to address such issues in 1991. In the electric utility field, disposal of nuclear waste and environmental concerns of electric power generation topped the agenda; 15 states expect to address such problems in 1991.

Another high-priority issue was to create incentives to encourage alternative power supplies. Cogeneration was a priority sub-issue in 23 states, while waste-to-energy conversion was a top sub-issue in 19. Increasing residential and commercial energy efficiency was selected as a high-priority issue by 50 percent of the states. Respondents showed a strong interest in statewide energy planning, renovation and weatherization of low-income housing, and statewide building codes. Also, creating financial incentives for implementation of conservation measures was a priority energy issue in nearly half the states. Another top issue was the use of alternative fuels for transportation. **Figure 6** displays the survey results on energy issues.

Figure 6

National Priority Ranking of Top Energy Issues

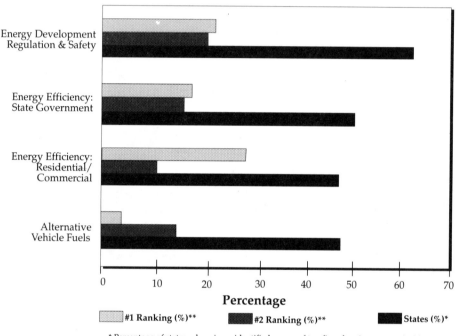

* Percentage of states where issue identified as one of top five electric energy priorities.
** Percentage of respondents who identified issue as a #1 or #2 electric energy priority.

Source: NCSL *State Issues 1991: A Survey of Priority Issues for State Legislatures, 1991.*

Similarly, in the utility areas, incentives to encourage customer energy conservation topped the agenda, with 23 states indicating this category as one of the top five issues for 1991. Least-cost planning and utility management were priorities for at least half the states, reflecting a growing need for IRP by utilities. IRP emphasizes a balanced approach to electrical demand management and a diversification of supply resources. **Figure 7** shows the survey results for utility issues.

Figure 7

National Priority Ranking of Top Utility Issues

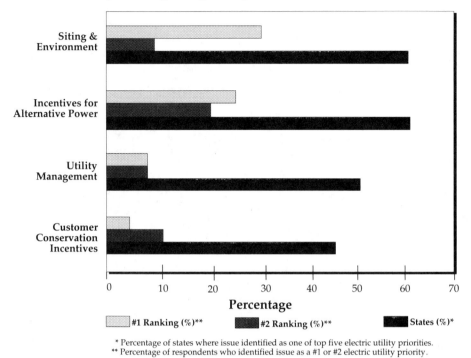

* Percentage of states where issue identified as one of top five electric utility priorities.
** Percentage of respondents who identified issue as a #1 or #2 electric utility priority.

Source: NCSL *State Issues 1991: A Survey of Priority Issues for State Legislatures, 1991.*

While many states consider new approaches to energy planning, state statutes, regulatory rules, or utility commission orders are already in place to promote IRP by regulated utilities in California, Massachusetts, Nevada, New York, Oregon, Vermont, Washington, and Wisconsin. California and Wisconsin are national leaders in encouraging electric utility demand-side resource investments. Respondents from these and other states expressed optimism that increased conservation and energy efficiency implemented by state governments can reduce the need for new electricity-generating capacity.

Despite the Persian Gulf crisis, energy issues played only a minor role in the November 1990 elections. However, there is a growing realization at the state level that energy is both scarce and expensive and that its use must be curtailed by conservation and increased efficiency standards. As state budget constraints increase, energy efficiency and conservation are expected to increase in importance in 1992 and beyond.

Benefits of Energy Efficiency and Conservation

States generally recognize that significant "spin-off" benefits can be realized through increased energy efficiency and conservation. One important benefit is a reduction in consumers' expenditures for imported fuel. This money can be put into savings or spent by consumers on other items that can benefit the state's economy. The average dollar value for one quad (1015 Btu or one quadrillion Btus) of energy savings in 2010 is estimated at $7 billion ($1984).[2] Once freed from its allocation to energy, this money could be spent on other goods and services to increase the sales revenues and employment levels of the businesses that supply these goods and services. Since a portion of the money would be put into savings, the supply of loanable funds would increase and stimulate new business investments in the states.

The building sector is an important target for energy-conservation measures by state legislatures, by means such as building standards, "energy-efficient mortgages,"* and least-cost energy planning.[3] The DOE provides an estimate of the approximate value of potential energy savings in the building sector in 2010.

Using price forecasts for 2010, **table 2** shows the potential savings in the buildings sector and their economic value for oil, natural gas, and coal as the fuel saved. **Table 2** also shows that the economic value of this energy savings depends on the type of fuel displaced.

Table 2
Savings Potential in the Residential and Commercial Building Sectors in 2010

Fuel	Residential		Commercial		Total	
	TBtu*	$1984/MBtu**	TBtu*	$1984/MBtu**	TBtu	$1984 (millions)
Petroleum	30.5	12.2	34.5	11.6	65.0	783
Natural Gas	98.6	10.9	83.4	10.5	182.0	1,947
Electricity	110.0	21.7	94.6	21.6	204.6	4,436
					Total	$7,166

*TBtu=Trillion Btus (10^{12} Btu)
**MBtu=Million Btus(10^6 Btu)

Source: *Energy Conservation Multi-Year Plan 1990-94*, U.S. DOE, Washington D.C., August 1988.

*Energy-efficient mortgage legislation requires that, prior to the sale of a building, economically justified energy-conservation measures be installed and the mortgage payments include repayment of the expenditures for these installations. Since the cost of the "saved energy" is less than the cost of energy for operating the building without conservation, the monthly payments for principal, interest, and utilities for the new owner are reduced and the mortgage amount for which the buyer qualifies increases.

A second benefit of reduced energy consumption is an improved balance of payments and decreased dependence on foreign sources of energy. A third benefit of conservation would be improved air and water quality. With reduced burning of fossil fuels, especially coal and petroleum, lower levels of air pollution would result. Although it is difficult to place a dollar value on environmental quality, reduced air-and water-pollution levels would cause fewer health problems and have positive aesthetic effects in terms of increased visibility and less thermal pollution. There would also be an improvement in the acid rain situation, particularly in the eastern and Great Lakes regions of the country. Acid rain, a result of the conversion of SO_2 and NOX into sulfuric and nitric acids, is creating dangerously acidic conditions in forest, agricultural, and lake areas. Motor vehicles and fossil-fueled power plants are the primary producers of SO_2 and NOX pollutants. Reduced pollution output could ameliorate the acid rain problem appreciably. Most state governments are aware of these potential conservation benefits and, as will be shown later in this book, have begun to take action to reduce energy consumption and pollution.

30

Chapter Notes

1   *NCSL State Issues 1991: A Survey of Priority Issues for State Legislatures* (Denver: NCSL, 1991).

2   U.S. Department of Energy, *Energy Conservation Multi-Year Plan 1990–94* (Washington, D.C.: 1988).

3   Ibid.

UTILITIES AND
ENERGY EFFICIENCY

*"Who lined himself with hope,
Eating the air on promise of supply."*

William Shakespeare, Henry the Fourth, Part I

More than one-third of the primary energy consumed in the United States is used to generate electricity. Thus, improved utility efficiency can reduce energy consumption appreciably. States play a key role because jurisdiction over electricity rates, utility planning, siting, certification, and system expansion is the responsibility of state public utility commissions (PUCs).

During the 1950s and 1960s, the low cost of primary energy sources and the economies of scale in electricity power production kept prices low. Low production costs led to high rates of return for utility stockholders. Regulators and utilities concentrated on the traditional function of utilities—to generate enough power to supply the demand. To estimate the growth in electricity demand they examined the historical relationships between GNP, population, fuel costs, and electricity consumption. Because financial rewards were largely based on how much supply could be generated, energy conservation and demand-side reduction were not considered in traditional planning. Utilities had "an obligation to serve" and concentrated on providing sufficient and reliable sources.

In the 1970s, fuel and power plant construction costs grew while the economy shrank. Utility stock prices and dividends plummeted and consumer electricity rates increased, bringing strong criticism from consumer interest groups and citizens. Increased internal costs led to greater cost consciousness by utilities, ratepayers, and regulators. Increased production efficiency and demand-side reduction were seen as ways to improve the earnings of utilities.

This spurred interest in IRP (also called least-cost planning [LCP]), which considers energy conservation measures on the same basis as supply options.[1]

IRP is a process to develop a long-term utility resource plan that accounts for both supply- and demand-side options for provision of future energy supplies at minimum cost and considers environmental and health concerns. Traditional supply-side options that increase the supply of electricity include life-extensions of existing power plants, construction of new fossil-fuel and nuclear power plants, and the upgrade of transmission equipment. Typical demand-side options that decrease the demand for electricity include such conservation measures as efficiency standards for appliances, industrial motors and lighting, improvements in heating and cooling of buildings, load management, and time-of-use rate incentives.

There are barriers to overcome before IRP becomes a successful policy tool.[2] IRP programs generally decrease electricity demand, which reduces sales revenues for utilities. Traditional rules and regulations have created disincentives for utilities to invest in energy efficiency because utilities in many states are rewarded financially based on how much electricity they generate and sell, regardless of actual demand. Unless utilities are rewarded appropriately for investments in conservation, they have no incentive to reduce demand. But a majority of utilities are beginning to realize that, with appropriate regulatory changes, IRP can help them achieve traditional goals without sacrificing profits. Using supply-and-demand options as future sources of electricity can reduce the need for new electric power plants, which require large capital investments and are risky in today's economic climate.

Utility Regulation and Incentives for Conservation

Traditional regulatory practices discourage utilities from conservation measures. Traditional rate structures allow profits for investor-owned utilities proportionate to their revenues— a reduction in electricity sales (by conservation) cuts revenues and reduces profit. However, regulators have begun to change traditional practices and look for ways to create energy-conservation incentives for utilities. A key to promote change is to uncouple utilities' earnings from the amount of power sold and to provide extra returns and tax credits for expenditures on energy-conservation measures that reduce customer bills. Some states reimburse utilities for the cost of efficiency investments; without rate structure change, this is inadequate to offset lower revenues to utilities resulting from successful demand-reduction programs. To correct this imbalance, a few coalitions of utilities, environmental groups, consumer groups, and public interest groups have collaborated with state commissions to develop profit incentives that help utilities earn a return on efficiency investments equal to that earned on traditional investments. The collaborative process has already resulted in implemented profit incentives in California, Massachusetts, New York, and Rhode Island.[3]

Traditionally, state utility commissions have allowed utilities to include reasonable profits in their rates based on reliable service to customers, adequate dividends to stockholders and interest to bondholders, and the costs to maintain and build plants and equipment. Part of the rate of return covered fixed costs, such as interest, and part covered a return on shareholders' equity. The latter was not guaranteed, but the utilities had ample opportunity to earn that return. Rates were supposed to reflect future expected costs.

This structure contains several counter-incentives to conservation. Because regulators allow only capital investments included in the rate base to earn a profit, it encourages utilities to build capital assets rather than contract for services. Regular expenses and contractual payments are recovered dollar for dollar out of rates; stockholders receive no profits on non-capital expenditures. Once rates are set, utilities can increase short-term earnings by selling more power than they had predicted, or by cutting expenses. Reducing demand by conservation measures only reduces profits. The incentive to increase supply by building more supply options reinforces the pressure to sell more power and justifies building more plants.

The situation is most apparent in states with excess capacity or more power to sell than customers need; as a remedy, utilities encourage demand, sometimes by offering rebates to replace gas with electric heating. Since no new facilities are needed to meet the new demand and profits are proportional to sales, these marginal sales are very profitable. Some utilities may now be able to recover costs for conservation measures and even make a small profit, but they cannot obtain the marginal profits that contribute substantially to the shareholders' earnings; to offset this disincentive some PUCs allow higher profits on capital investment in conservation measures. This approach can create a problem in a competitive market if electricity costs become too high and alternative sources are available. It may also encourage utilities to invest in expensive or inefficient conservation measures when the amount of extra return earned is simply a function of the amount of capital invested. The general consensus of state legislators is that the time is ripe for regulatory practices to create incentives for conservation that will reduce the need for new electric power sources.

State Actions

Several states have innovative ways to treat "conservation-lost revenues," but none has been without problems. Approaches taken by five state utility commissions follow.

California. The California PUC introduced an electric revenue adjustment mechanism (ERAM) that attempts to separate revenues from power sales. This part of the state's rate structure compares a utility's actual and predicted revenues and adjusts rates in succeeding periods based on the amount above or below the forecast. This permits utilities whose revenues decrease as a result of a successful conservation program to recoup their losses by raising electric rates in subsequent periods. ERAM decouples profits from electric power sales and protects a utility's earning from drops in demand.

Idaho. The Idaho PUC recently supported utilities' turning to conservation for new sources of power. The order says the commission will allow utilities higher rates of return on efforts that lead to energy conservation. That includes so-called "lost-opportunity resources"—energy savings that utilities achieve by encouraging builders to construct energy-efficient new homes and offices, for example. In the past, Idaho allowed utilities to receive a return on some conservation expenditures—allowing them, over 20 or 30 years, to recover money spent plus interest, and to recover in one year other investments dollar for dollar. The new order allows utilities a return on all conservation investments.

Montana. State regulations allow the Public Service Commission (PSC) to grant 2-percent higher returns on conservation investments than on other utility investments. It also allows utilities to earn a return on their conservation costs through adjusted rates over 30 years.

Oregon. The Oregon PUC allows utilities to earn a return on approved conservation expenditures. Some generalized conservation spending, such as administration costs, are not allowed to earn a return.

Washington. The state Utilities and Transportation Commission allows utilities to earn a return on conservation expenditures and permits one company—Puget Sound Power and Light—to adjust rates automatically between rate hearings. It also allows utilities to earn a 2-percent higher return on conservation expenditures than on other resource spending.

Because conservation is often the cheapest and environmentally most benign energy resource, regulatory changes such as these can play a leading role in meeting the states' electric power needs at the lowest cost. They are relatively easy to implement, they are abundant, and their effect can grow as demand for power increases.

Projecting future electricity generating capacity needed to meet U.S. demand is highly uncertain. As **figure 8** indicates, different projections of electricity demand in the United States through 2010 vary by approximately 50 percent.[4] By diversifying into the demand-side energy efficiency, utilities become more flexible and able to meet the needs of electricity-demand forecasting. It is expected that IRP and energy conservation will dominate state energy agendas in the next few years. Through the DOE, the federal government has pledged financial and technical support for IRP.

Figure 8
Projections of United States Demand for Electricity

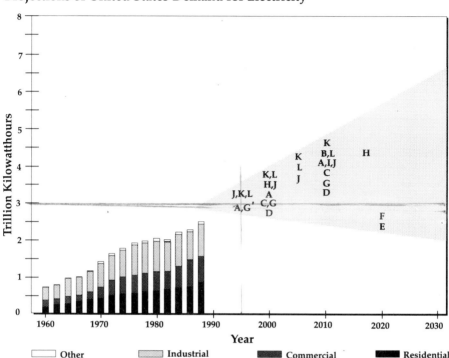

A Gas Research Institute, 1991 GRI Baseline.
B American Gas Association, Total Energy Resource Analysis Model, 1989.
C Commission of the European Communities, Directorate, General for Energy. *Energy in Europe, MajorThemes In Energy, 1989.*
D Commission of the European Communities, 1989 ("Cost-Effective Conservtion" case).
E J. Goldemberg et al., *Energy for a Sustainable World,* 1988 (high).
F Goldemberg, 1988 (low).
G Data Resources Inc./McGraw-Hill, Energy Review, 1990.
H Edison Electric Institute, *Electricity Futures: America's Economic Imperatives 1989,* 1989.
I WEFA Group, *Energy Analysis Quarterly, Long-Term Economic Outlook,* 1989.
J Energy Information Administration (EIA), Annual Energy Outlook, 1990 (low growth case).
K EIA, 1990 (high growth case).
L EIA, 1990 (base case).

Notes: Differences in projections are caused, in part, by varying assumptions concerning energy prices, economic growth, consumer and producer behavior, and rates of technological change, including replacement of capital stock. The shaded area represents an envelope bracketing these differences.

Sources for E & F are very questionable. General conclusions and data indicate that electric use is proportional to GNP.

Source: U.S. Department of Energy, *The National Energy Strategy,* 1991.

Chapter Notes

1   L. Berry and E. Hirst, "The U.S. DOE Least-Cost Utility Planning Program," *Energy* 15 (1990): 1107–1117.

2   D.H. Moskovitz, "Cutting the Nation's Electric Bill," *Issues in Science and Technology* (Spring 1989): 88–93.

3   M. Deland and J. Watkins, "Perspectives on Electric Efficiency," *Issues in Science and Technology* (Winter 1990–1991): 32.

4   U.S. Department of Energy, *The National Energy Strategy* (Washington, D.C.: February 1991).

INTEGRATED RESOURCE PLANNING

"To what purpose is this waste?"

Matthew 26:8

Integrated resource planning is the process that simultaneously examines all energy-saving/producing options to optimize resources and minimize total consumer costs; its elements include consideration of environmental and health concerns. **Table 3** shows that at least 23 states already use an IRP framework for future electricity planning and another 19 are developing, implementing, or considering IRP frameworks.[1]

Table 3
Current Status of Least-cost Planning (LCP) Implementation

LCP in Practice (23)		Implementing LCP (8)	Developing/ Considering (11)	Not Considering/ Not an Issue (9)
Arizona	Nevada	Hawaii	Alaska	Alabama
California	New Hampshire	Michigan	Arkansas	Louisiana
Connecticut	New York	Minnesota	Colorado	Mississippi
Delaware	North Carolina	New Jersey	Georgia	Nebraska
Florida	Ohio	North Dakota	Idaho	New Mexico
Illinois	Oregon	Rhode Island	Kansas	South Dakota
Indiana	Pennsylvania	South Carolina	Kentucky	Tennessee
Iowa	Vermont	Texas	Missouri	West Virginia
Maine	Virginia		Montana	Wyoming
Maryland	Washington		Oklahoma	
Massachusetts	Wisconsin		Utah	

Sources: EPRI, *Least-Cost Planning in the United States*: 1990 (CU-6866), Palo Alto: 1990; EPRI, *Demand-Side Management: Utility Options for the Future*, Palo Alto: 1989.

38

Figure 9
Schematic Diagram of IRP Process

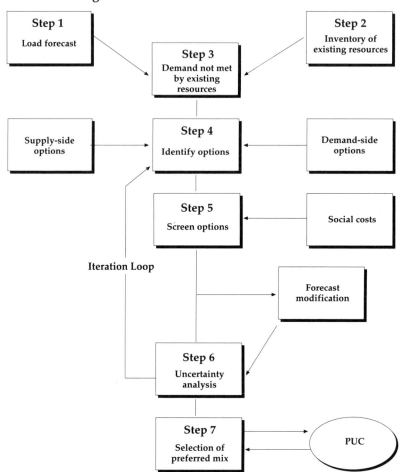

There is no unique method for IRP. Nearly every state government, PUC, or utility adds its own touch to the process. An extensive study of IRP development procedures in the United States, conducted by Oak Ridge National Laboratory,[2] showed that the following seven steps are generally used: development of a load forecast; inventory of existing resources; identification of future electricity needs that will not be met by existing resources; identification of potential resource options, including demand-side reduction programs; screening of options to identify those that are feasible and economic; performance of some form of uncertainty analysis; and selection of a preferred mix of resources. Although not all of these elements are always used in the order listed, most resource plans are developed by some combination of the above steps.

Figure 9 shows a model IRP process used to develop a procedure to evaluate demand-side options (conservation and load management), compare these to supply-side options, and structure an energy policy that integrates environmental and social costs to develop a long-term energy policy that acquires the most inexpensive resources first and internalizes social costs in the rate structure.

The diagram can be viewed as a series of top-down steps. The conventional supply-side issues feed from the left; the measures and issues arising from demand-side options and social cost considerations feed the right. In view of the uncertainty about future load growth, fuel prices, capital costs, and consumer response to demand-side programs, a "sensitivity analysis" is incorporated. The feedback loop allows examination of various scenarios based on different assumptions for the parameters. Under such a sensitivity analysis, the "most likely" plan—based on a combination of options using average predictions—is examined first. Key factors are then varied to determine how the initial plan responds to these variations. Modifications can then be made for another iteration. In subsequent sections, key elements of the demand-side issues will be considered further.

Demand-side Management

Demand-side management (DSM) is a broad term that encompasses planning, implementation, and evaluation of utility-sponsored programs to influence the amount or timing of customers' energy use. This in turn affects the system energy (kWh) and total capacity (kW) that the electrical utility must provide to meet the demand. DSM is a resource option that complements power supply. It is used to reshape customer energy use and demand, thus providing an important component of a modern utility's energy resource mix.

Four basic techniques influence energy use in DSM. They are shown schematically in **figure 10** and described below.

Peak Clipping is the reduction of the system peak loads. It is a classic form of load management. Peak clipping generally reduces peak load by using direct load control, commonly practiced by utility control of customer appliances. While many utilities use this mainly to reduce peaking capacity or capacity purchases during the most probable days of system peak, direct load control can also be used to reduce operating cost and dependence on critical fuels.

Valley Filling builds off-peak loads. This may be particularly desirable where the long-run incremental cost is less than the average price of electricity because adding lower priced, off-peak load under those circumstances decreases the average price. One of the most popular valley filling methods is to use thermal energy storage for water heating and/or space heating.

Figure 10
Demand-side Management Techniques

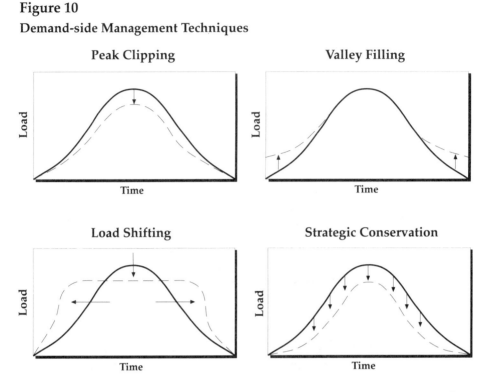

Load Shifting shifts load from on-peak to off-peak periods. Examples include use of storage water heating, storage space heating, cold storage, and customer load shifts. The load shift from storage devices displaces loads which would have existed if conventional appliances without storage had been installed by the customer.

Strategic Conservation is the load shape change that results from utility-stimulated programs directed at reducing end-use consumption. It was not always considered load management because reduction of the load involves a reduction in sales as well as a change in the pattern of use. In employing energy conservation, the utility planner must consider what conservation actions would occur naturally and evaluate the cost-effectiveness of possible utility programs to accelerate or stimulate those actions. Examples include weatherization and appliance efficiency improvement.

Table 4 shows how various conservation technologies in the residential and industrial sectors affect load management. For example, improved insulation for a building reduces energy consumption and is therefore classified as strategic conservation. Using a high-efficiency compressor in an air conditioning system reduces consumption during peak load and therefore achieves peak clipping as well as strategic conservation.

In addition to the four energy-conservation and load-shaping programs used by utilities to influence the amount or timing of customer energy use, utilities with excess capacity often attempt to increase their sale of power. This process is called strategic load growth and may involve

Table 4
Load Impacts of Conservation Technologies

	LOADSHADE IMPACTS			
TECHNOLOGY	PEAK CLIPPING	VALLEY FILLING	LOAD SHIFTING	STRATEGIC CONSERVATION
Building Envelope				
Roof and wall insulation				X
Double glazing				X
Vestibules				X
Solar films	X			X
Infiltration/exfiltration				X
Lighting				
Low wattage incandescent	X			X
HID lighting	X	X		X
Low wattage fluorescent	X			X
High efficiency ballasts	X			X
Compact fluorescents	X			X
Daylighting controls	X			X
Reflectors	X			X
Heating/Ventilation/Motors				
Destratification fans				X
Heat pumps				X
Gas absorption chillers	X	X		X
Economizers				X
High efficiency gas furnace				X
Exhaust air heat recovery				X
High efficiency chillers	X			X
High efficiency motors	X			X
Optimum motor sizing	X			X
Reduce ventilation	X			X

Table 4 (continued)
Load Impacts of Conservation Technologies

	LOADSHADE IMPACTS			
TECHNOLOGY	**PEAK CLIPPING**	**VALLEY FILLING**	**LOAD SHIFTING**	**STRATEGIC CONSERVATION**
Appliances and Water Heating				
High efficiency A/C	X			X
Heat recovery water heater				X
Tank insulation				X
Solar water heater	X			X
Low temperature dishwasher	X			X
Efficient cooking equipment	X			X
Load Management				
Pool pump controls	X		X	
Water heater cycle controls		X	X	
Timers			X	X
Radio controlled central A/C	X		X	
Thermal storage	X	X	X	
Programmable thermostats				X
Refrigeration				
Glass/acrylic doors				X
High efficiency compressors	X			X
Automatic defrost controls	X			X
Condenser heat recovery				X
Pipe insulation				X
Tank insulation				X

Source: NCSL, 1992.

incentives to switch from gas to electric appliances or rebates for the installation of electric heating and cooling systems. New electric technologies for industrial process heating, transportation with electric trains or vehicles, air conditioning of buildings, and so forth, is called electrification. These measures increase the primary energy consumption and environmental costs, but may reduce the use of fossil fuels if nuclear power is used. The use of renewable and solar technologies could eventually allow the growth of electric power with small environmental impacts.

Table 5 shows the economic cost and payback time for some typical conservation measures as compiled by the Solar Energy Research Institute (SERI) for the Western Area Power Administration (WAPA).[3] The cost of the equipment and installation in column one, the energy savings per year in column two, the yearly cost savings in column three, the simple payback in column 4, and the life in column five were taken from note 50. The cost of saved energy in the last column of **table 5** was calculated from the report, using a real discount rate of 3 percent. Also, a maximum life of any energy savings technology of 20 years was used in these calculations.

Table 5
Estimated Cost and Payback of Energy-saving Technologies

Technology	Installed Cost $	Energy Savings (kWh/yr)	Cost Savings ($/yr)	Simple Payback (yr)	Life (yr)	Cost of Saved Energy (c/kWh)
Building Envelope:						
Insulation						
Wall	800	2831	226	1.2-4.6	50	1.90
Ceiling	1140	4916	393	1-3.8	50	1.56
Foundation	648	2880	230	2.1-8.4	50	1.51
Windows, Double Pane, Low-E, N	2.75/SF	6.3/SF	0.51/SF	5.4	20	2.93
Windows, Double Pane Low-E, S	2.75/SF	4.5/SF	.40/SF	6.9	20	3.42
Glass Storm Window	5/SF	9.5/SF	.76/SF	4.6-10.6	20	3.54
Solar Films	1.85/SF	4.75/SF	.38/SF	4.8	3-15	3.26-13.78
Weatherstripping/ Caulking	230	1852	148	1.6	2.5	5.23
HVAC/Motors						
Repair Duct Leaks	110	2300-4000	184-320	0.3-0.6	7	0.61
Duct Insulation	0.85/LF	2-7.5/LF	.16-0.58/LF	1.5-5.3	25	1.55
Heat Pumps						
Air-source, Cold Climate	4130	3190	255	4.315	2.92	
Air-source, Hot Climate	3920	3884	311	2.9	15	1.94
Destratification Fans	415	2666	213	1.9	10	1.82
Efficient Air Conditioners	300/ton	600/ton	48/ton	6.2	15	4.19
Heat Exchangers	3760/kcfm	17000/kcfm	1190/kcfm	3.1	20	1.49
Direct Evaporative Cooling	850/ton	1241/ton	87/ton	2.35-20	1.08-3.51	
EMCS	300	2790	195	1.5	10	1.26
Economizer	62.5	162-1785	11.5-125	0.5-5.5	15	2.38
Efficient Motors (1-5hp)	166	520	-	1.3	7	5.12
Optimum Motor Sizing	192	-	53	3.6	7	-

Table 5 (continued)
Estimated Cost and Payback of Energy-saving Technologies

Technology	Installed Cost $	Energy Savings (kWh/yr)	Cost Savings ($/yr)	Simple Payback	Life (yr)	Cost of Saved Energy (c/kWh)
Appliances and Water Heating						
Heater Wrap	21	273	19.11	1.1	10	0.90
Thermal Traps	8	380	30	0.3	15	0.25
Pipe Wrap	5	20	1.60	3.1	10	2.93
Low-flow Shower Head	9	275	22	0.4	10	0.38
Heat-pump Water Heater	1350	2780	222	4.7	13	3.50
Heat-recovery Water Heater	700	1100	88	7.9	13	5.98
Solar Hot Water Low Flow System	42/SF	81/SF	6.50/SF	6.5	20	3.49
Refrigerators and Freezers	731	590	47.2	1.3	20	0.68
Low-water Washing Machine	150	480	38.4	3.9	15	2.62
Cook Tops	302	44	3.52	0.6	18	0.33
Ovens	215	103	8.24	1.8	18	1.06
Lighting						
Efficient Incandescent	1.13	8	0.48	469 (hr.)	750 (hr)	6.84
Compact Fluorescent	13	57	5.67	2150 (hr)	10000 (hr)	2.44
High Efficiency Fluorescent	18	153	9.3	0.9	20000 (hr)	2.42
Exit Light Conversion Kits	39	289	29.36	1.32	10000 (hr)	2.67
Photocell	12	219	17.52	0.7	5	1.20

Low-E = Low emissivity coating on glass
SF = Square foot
LF = Lineal foot
kcfm = 1000 cubic feet per minute

EMCS = Energy management and control systems
hr = Hour
HVAC = Heating, ventilation, and air conditioning

Source: WAPA, 1991. (See reference 3 at the end of this chapter.)

For comparison, **table 6** shows the cost of new power generation estimated by the California Energy Commission.[4] Note that even though the cost of new electrical power from various sources does not include any social cost, the cost of saved energy for a majority of the conservation measures in **table 5** is considerably less than the cost of power from new generating sources, including nuclear and coal.

Table 6
Estimated* Cost of New Electrical Generation in Constant 1987 Dollars[5]

	Low	High
	(in cents/kWh)	
Solar Thermal Hybrid	6.0	7.8
Nuclear	5.3	9.3
Natural Gas (Intermediate)	5.3	7.5
Hydro	5.2	18.9
Wind	4.7	7.2
Coal Boiler	4.5	7.0
Natural Gas Combined Cycle	4.4	5.0
Geothermal Flash Steam	4.3	6.8
Biomass Combustion	4.2	7.9

*Note: These estimates do not include social costs.

Source: California Energy Commission, 1990.

LCP and DSM for Gas Utilities

Gas utilities have only recently begun to use LCP and DSM. Previously, gas utilities were mainly concerned with selling fuel, not with comprehensive long-range planning. But with natural gas targeted by the NES for increased use and considered by some—Senator Timothy Wirth (Colo.), for example—as a "bridging fuel" to replace imported oil, gas utilities have begun to review their future plans, options, supplies, and pipeline capacity limitations.

A recent overview of these efforts by four western gas utilities showed that their planning process is quite similar to that of the electric utilities.[6] However, there is a difference between the options available to utilities that provide both gas and electricity and those that provide gas only. The former have the option to encourage fuel switching, that is, to provide incentives for electric customers to convert their electric space and water heating equipment to natural gas. This shift would promote an increase in energy efficiency, reduce total utility energy cost, and decrease the detrimental environmental impact. DSM options open to both include incentives for weatherization of buildings, replacement of old furnaces with high-efficiency models, installation of electronic ignitions and automatic vent dampers, and installation of low-flow shower heads.

The main difference between gas-only utilities and those that provide both gas and electricity is that the former have only limited control over the nature and timing of large resource additions, and may be forced to cease operation when their gas supply runs out. Demand-side reduction options are similar for both types of utilities. However, since the fuel cost for gas-fired appliances and gas-heated buildings is much less than for electrical supply, DSM programs that are technically feasible may not be economically viable for gas-only utilities. Hence, demand-side resources tend to be limited within a planning horizon of 10 to 15 years.

In June 1991, the National Association of Utility Regulatory Commissioners (NARUC) Committee on Energy Conservation and LBL jointly conducted a survey of LCP by gas utilities.[7] Their findings for five western states are summarized below.

California (under development). Many of the ingredients of a natural gas utility LCP process are "fairly well developed," although there is no formal or regular proceeding concerning gas LCP. Long-range gas supply- and demand-side options are included in the California Energy Commission's *Biannual Fuels Report*.

Hawaii (under development). The PUC is currently involved in a proceeding to establish a LCP framework for electric utilities and GASCO, a local synthetic gas producer and distributor.

Nevada (in practice). Gas LCP requirements were developed as a result of electric LCP/IRP requirements and a 1987 legislative mandate. The Southern Division of Southwest Gas Corporation filed its LCP on July 1, 1990. The DSM portion was rejected by the PSC. The Northern Division must file by January 1992.

Oregon (implemented). In April 1989, the Oregon PUC implemented electric and natural gas IRP after a formal investigation. Northwest Natural Gas and Cascade Natural Gas have submitted IRPs to the commission.

Washington (implemented). Based on a 1987 order, Washington's gas utilities must prepare a LCP in consultation with commission staff. One gas utility, Washington Water Power, has submitted a LCP to the Washington Utilities and Transportation Commission.

Montana and **Colorado** are considering gas utility LCP regulations.

In the nation, 15 states are actively developing gas utility LCPs, while 29 others are not pursuing gas LCP. In addition to examining regulatory commission gas LCP activities, the survey looked at DSM programs for gas utilities. Eighteen states allow gas utilities to develop gas conservation programs on a voluntary basis, while others are experimenting with incentives and regulations to encourage gas utility conservation programs.

IRP Models

Stone and Webster Management Consultants, Inc., has compiled and analyzed a list of 47 IRP models for the WAPA.[8] According to this overview, computer models for IRP have been developed by a number of organizations and companies, including SERI (now NREL), the Canadian Electrical Association, Lawrence Berkeley Laboratory (LBL), the California Energy Commission, the Electric Power Research Institute (EPRI), the American Public Power Association (APPA), the National Rural Electric Cooperative Association (NRECA), and DOD.

Most of these models are complex and require dedicated personnel and hardware, as well as extensive training, to operate. They are designed for use by large utilities, with licensing fees ranging from $10,000 to $300,000. These models include COMMEND (developed by EPRI), SRC/COMPASS (developed by Synergic Resource Corporation), DSM ASSYST (developed by XENERGY), and UPLAN (developed by Lotus Consulting Group).

Several smaller and more manageable models are available for use as planning tools by small utilities, public utility commissions and state agencies. Examples include COMPLEAT (developed by APPA), CPAM (developed by the University of Southern California), DSSTATEGIST (developed by Barakat, Howard & Chamberlain), and SMART ENERGY (developed by Synergic Resource Corp. for APPA). Each of the available models has varying capabilities and design strategies. For instance, COMMEND is "designed to forecast commercial customer energy consumption for 12 building types, eight end uses and five fuel types," while CPAM analyzes conservation strategies by end use type and projects' "likely energy savings and the cost of the [conservation] program over time." Stone and Webster Management Consultants, Inc., is currently planning to develop a versatile IRP model, designed specifically for small utilities in the WAPA, if demand justifies this development.[9]

Supply Curves of Conserved Energy

A useful tool in IRP is the supply curve of conserved energy. It portrays graphically the technical potential of various conservation options and their energy costs. Appendix 3 presents a procedure developed at LBL for constructing a supply curve of conserved electricity in a consistent assumptions framework.

Chapter Notes

1   *Least-Cost Planning in the United States: 1990* (CU-6966, Research Project 2982–2, prepared for EPRI by Barakat & Chamberlain, Inc., Palo Alto: 1990).

2   M. Schweitzer et al., *Key Issues in Electric Utility Integrated Planning: Findings from a Nationwide Study* (ORNL/CON-300, Oak Ridge, Tenn.: April 1990).

3   SERI, *Demand Side Management Pocket Guide Book, Volume 1: Residential Technologies* and *Volume 2: Commercial Technologies.* Prepared for the Energy Services. (Golden, Colo.: SERI, April 1991).

4   California Energy Commission, 1990.

5   Ibid.

6   Washington Water Power, Northwest Natural Gas, Southwest Gas Corporation, and Southern California Gas, "Least-Cost Planning and Demand-Side Management for Gas Utilities," 1991.

7   NARUC Committee on Energy Conservation and LBL, "Survey of LCP by Gas Utilities," 1991.

8   Stone and Webster Management Consultants, Inc., "Model Review, Phase 3–WIRP," prepared for the Western Area Power Administration, April 1991.

9   C. Council of WAPA, 1991: personal interview.

SOCIAL COSTS OF ENERGY

"It is as fatal as it is cowardly to blink facts
because they are not to our taste"

John Tyndall

tate agencies that regulate electric utilities, resource acquisition rates,
and utility planning have recently come under pressure to include
the costs of "externalities," usually called the "social costs," in their
resource planning. Externalities occur when the producer does not pay for all
the costs of production. Air pollution from utilities is a typical example of an
externality cost (EC) that is currently not included in the production cost and
becomes a social cost. The concept of including externalities in electric power
planning originated in the Pacific Northwest Power Planning and
Conservation Act of 1980, which specified a cost-effectiveness criterion for
resource evaluation that includes the quantification of environmental costs and
benefits. Since then, 26 states have begun requiring public utilities to
incorporate external and environmental costs in utility resource planning. A
1990 survey by LBL for the NARUC showed that agencies in 17 states have
already developed approaches to incorporate certain environmental costs in
their resource planning.[1] Some of these state efforts are discussed below.
Details on the programs for six key states are provided in part two of this
book.

Results of Previous Studies

Several analysts have attempted to quantify the social costs of
pollution associated with various energy systems and a few regulatory bodies
have made preliminary efforts to incorporate ECs into the resource planning
process.[2] **Table 7** shows a summary of previous EC studies compiled by
Koomey.[3] ECs are difficult to estimate because they depend on assumptions

Table 7
Overview of Externality Studies

Study	Insult/Stress	Method	Locale	Remarks
1. Shuman and Cavanaugh (Cavanaugh et al., 1982	Human health, property damage, CO_2 - Property damage, famine	DE	Northwest U.S.	Attempt to quantify difficult damage effects from global warming
2. EPRI (1989)	SO_2	DE	Rural PA and WV and 35 mi. from NYC	Koomey's analysis uses EPRI's mid-range rural estimate
	NOX	DE	Not specified	
3. Hohmeyer (1988)	CO_2, NOX, SO_2 ROG, particulates, subsidies, nuclear, accidents, depletion	DE	West Germany	Marginal damage costs per pound specified in a personal communication (1990)
4. Chernick and Caverhill (1989)*	Particulates, oil imports	DE	New England	
	SO_2, NOX, CO_2, CH4, oil spills	AC and DE	New England	
5. Schilberg (1989)	SO_2, NOX, CO_2 ROG, CH4, N2P	AC	Outside CA, CA inside and outside SCAQMD	
6. CEC Staff (1989)	SO_2, NOX, CO_2, particulates, ROG	AC	California	From 1990 Electricity Report
7. NPPC (1989)	Financial risks from lack of modularity	RD	Northwest	10 percent adder when comparing supply resources to energy conservation
8. NYPSC (1989b)	SO_2, NOX, CO_2, particulates, water, land use	AC and RD	NY state	Con Ed bidding system (NY City)
9. PSCWI (1989)	All	RD	Wisconsin	Non-combustion credit, cost reduced by 15 percent in resource planning
10. NRDC (1989)	CO_2	Other	California	Proposal - builds on NPPC with variable percent adders for CO_2

Method refers to direct estimation of costs (DE), abatement costs (AC), or regulatory determination (RD).
ROG = Reactive organic gases: SO_2 - sulfur dioxide, NOX = nitrogen oxides, CO_2 = carbon dioxide, CH4 = methane.
*Chernick and Caverhill rely mostly on abatement costs and this analysis uses their abatement cost estimates for SO_2, NOX, and CO_2.

Source: J. Koomey, *Comparative Analysis of Monetary Estimates of External Environmental Costs Associated with Combustion of Fossil Fuels*, Lawrence Berkeley Laboratory, Berkeley: July 1990.

and value judgments. There is, however, general agreement that the use of energy sources, as well as conservation measures, generate social costs not reflected in market transactions. According to Griffin and Steele, "external costs exist when the private calculation of benefits or costs differs from society's evaluations of benefits or costs."[4]

Table 8
Detrimental Environmental and Health Impacts
From Fossil Fuels and Nuclear Power (Incomplete)

Activity	All Fossil Fuels	Natural Gas and Oil	Coal	Nuclear
Exploration/ Harvesting	Air pollution (CO_2 CH4, N_2O, NOX, CO, ROG, HCs, SO_2) trace metals, particulates, thermal pollution, global warming	Drilling accidents, drilling sludge disposal	Mining injuries, acid mine drainage	Mining injuries
Processing/ Refining	Air pollution (CO_2, CH4, N_2O, NOX, CO, ROG, HCs, SO_2) trace metals, particulates, thermal pollution, global warming	Refinery accidents, refinery waste disposal	Respiratory disease	Hazardous chemical wastes
Transpor- tation/ Distribution	Air pollution (CO_2, CH4, N_2O, NOX, CO, ROG, HCs, SO_2) trace metals, particulates, thermal pollution, global warming	Pipeline and refrigerated tanker accidents, LNG explosions	Train accidents	Accidental release of radiation possible
Storage/ End Use	Air pollution (CO_2, CH4, N_2O, NOX, CO, ROG, HCs, SO_2) trace metals, particulates, thermal pollution, global warming	Storage accidents	Acid disposal, acid rain	Radioisotope emission, increased thermal pollution, long-term storage of radiation products may be health hazard

CO_2 = carbon dioxide
CH4 = methane
N_2O = nitrous oxide
NOX = nitrogen oxides
LNG = liquid nitrogen gas

CO = carbon monoxide
ROG = reactive organic gases
HC = hydrocarbon
SO_2 = sulfur dioxide

Source: NCSL, 1991.

Pollution, land use, global warming, and adverse health effects represent external costs because damages associated with those factors are borne by society as a whole and are not reflected in market price. **Table 8** presents a summary of key detrimental environmental impacts from fossil fuel use that result eventually in social costs.

There are at least two ways to calculate ECs. One uses direct damage estimates, which involves calculating damages that can be causally linked by pathways to emissions of a particular pollutant in economic terms.[5] In one such study, the human health and environmental effects due to coal consumption in new power plants was estimated.[6] This work illustrates that direct estimation of social cost is extremely complicated because many of the pathways are difficult to quantify while others are not fully understood.

The other approach determines the cost of abatement by estimating the cost of pollution controls imposed by regulatory decisions as a proxy for the true ECs of pollution. This approach assumes that regulators' choices embody societal preferences for pollution control, that the marginal costs of mitigation are known, and that these costs are incurred solely to reduce emission of a given pollutant. The last of these assumptions ignores the fact that auxiliary benefits sometimes result from investment in pollution reduction, such as when energy conservation reduces pollution and respiratory diseases. Problems and pitfalls involved in calculating societal costs associated with energy technologies have been studied by Holdren. Apart from the questions of precision, averaging and drawing appropriate boundaries, Holdren believes that results can be biased because analysts often focus on those things that are amenable to quantitative treatment, but ignore items that are difficult to quantify. He identifies this problem as "confusing things that are countable with what counts."[7]

Cost of Externalities and Cost (or Value) of Energy Savings

EC is the total cost in dollars to be borne by society because it is not part of the market cost. The EC of an adverse environmental or health impact (AEI) in dollars can be estimated from the following relation:

Externality Cost = Size of Impact X Damage Cost per Unit of Impact

Size of the adverse environmental or health impact is expressed in physical units, such as pounds of pollutants emitted from a power plant or number of people that contract respiratory diseases as a result of the emission from a power plant. Damage cost per unit of impact is its economic effect in dollars per unit of AEI, such as average cost of respiratory disease cure per person.

ECs must be based on a common unit of service, usually the delivered kWh (including transmission and distribution losses) for electricity or the delivered MBtu for direct fuel consumption (such as, natural gas combustion).

The cost (or value) of conserved energy (CCE) is equal to the initial investment cost of the conservation device in dollars times the capital recovery factor (CRF) divided by the annual amount of energy saved, or

$$CCE = \frac{\text{Cost of Conservation Device X CRF}}{\text{Energy Saved Per Year}} \quad \text{(1)}$$

The capital recovery factor accounts for the time value of the money invested in the device and its numerical value depends on the lifetime of the conservation device, t, and the discount rate, r, or

$$CRF\ (r,t) = \frac{1}{r}\left[1 - (1+r)^{-t}\right] \quad \text{(2)}$$

For example, if the discount rate is 3 percent and the lifetime is 10 years

$$CRF\ (3, 10) = \frac{1}{0.03}\left[1 - (1.03)^{-10}\right] = 8.53 \quad \text{(3)}$$

For short payback or lifetimes and low real discount rates (typically less than five years and below 3 percent) as a first approximation, the capital recovery factor may be taken as equal to the reciprocal of the lifetime. For accurate analysis, additional factors such as fuel escalation, inflation and tax rates must be considered, as shown by Kreith and Kreider.[8]

A real discount rate widely used for estimating energy savings is equal to the interest rate charged by a bank minus the inflation rate. A value of 3 percent is a reasonable number, according to Goldstein et al.[9] The following examples use a CRF equal to the reciprocal of the life of the device.

$$\frac{\text{Cost of Energy Saved}}{\text{(in \$/Btu or \$/kWh)}} = \frac{\text{Cost of Device in Dollars (\$)}}{\text{Lifetime (yr) X Energy Saved Per Year (Btu/yr or kWh/yr)}}$$

For example, suppose that installing an insulating blanket on a domestic hot water heater saves 6 MBtu/year. If the cost of the blanket is \$30 installed and the life of the blanket is the same as the guarantee on the water heater, five years, the cost (or value) of the conserved energy is:

$$CCE = \frac{\$30}{(6 \text{ X } 10^6 \text{ Btu/yr}) (5 \text{ yr})} = \$1/\text{MBtu}$$

By comparison, the cost of natural gas entering the house is about $3/MBtu. If the combustion efficiency is 75 percent, the delivered energy cost is $4/MBtu.

Hence, the cost of conserved energy for the gas-heated system is $1/MBtu, which is (1:4) x 100 or 25 percent of the fossil fuel cost. But if this domestic water heater were electrically heated and the electricity cost were 5.5 c/kWh (1 kWh = 3,412 Btu), the comparison would show an appreciably greater economic savings. Whereas the cost of gas heat delivered was $4/MBtu, the cost of electric heat is:

$$CCE = 0.055 \ \frac{\$}{kWh} \ X \ \frac{1kWh}{0.003412 \ MBtu} \ = \$16/MBtu$$

Hence, the conserved energy cost in this case is about (1:16) x 100, or 6 percent of the cost of the extra electric power needed if the heater were not insulated.

The social costs of energy can be estimated from an economic perspective if the size of impact and its unit cost are known. External environmental and health costs of energy have also been estimated grossly by means of regulatory determinations designed to include environmental impacts in regional LCP. For example, the Northwest Power Planning Council initiated in 1980 a planning process that gives conservation a 10-percent cost advantage over conventional supply resources. The Vermont Public Service Board instituted in April 1990 a 10-percent comparative risk and flexibility "adder" and a 5-percent environmental externality adder for comparing energy conservation with avoided costs of supply technologies. The Public Service Commission of Wisconsin implemented in 1989 a "non-combustion credit" of 15 percent for non-fossil supply- and demand-side resources excluding nuclear power, which is restricted by law in the state. The Wisconsin approach reduces conservation and renewable costs by 15 percent so that a conservation measure costing $100 per kilowatt is treated as though it costs only $85 per kilowatt in Wisconsin's resource planning process. This policy is equivalent to increasing the cost of conventional fuel resources by 17.6 percent.

The principal advantage of technology-based percentage adders is their simplicity and ease of implementation. However, according to Koomey, this approach submerges important details for specific projects of the same technology type.[10] For example, pollutant levels for a given technology may differ significantly, depending on the project size and design characteristics. Moreover, the economic effect of a given project may depend on its location; for example, air pollution in a sparsely populated area may be less significant and cheaper to remedy than in a densely populated area such as New York or Los Angeles. This is illustrated by the estimates for EC from new power plants in **table 9**.

Table 9
Externality Costs from New NSPS Power Plants

Study	Method	Location	Pollutant	Average delivered cost (cents/kWh) 5.5 Natural Gas - CT[+]	8.3 Coal-Steam Turbine[+]
1. EPRI (1987)	DE	Rural PA and WI; and 35 mi. from NYC	SO_2 + NOX (Range)	0.006 - 0.068 {0.11%–1.2%}	0.14 - 0.67 {1.7%– 8.1%}
2. Hohmeyer (1988)	DE	West Germany	SO_2 + NOX (Low)	0.09 {1.6%}	0.32 {3.9%}
			SO_2 + NOX (High)	0.46 {8.4%}	1.72 {21%}
3. Chernick and Caverhill (1989)	AC&DE	New England	SO_2 + NOX + CO_2	2.32 {42%}	3.98 {48%}
4. Schilberg (1989)	AC	Outside CA	SO_2 + NOX + CO_2	1.59 {29%}	2.71 {33%}
		CA Outside SCAQMD	SO_2 + NOX + CO_2	3.96 {72%}	7.90 {95%}
		CA Outside SCAQMD	SO_2 + NOX + CO_2	4.80 {87%}	14.72 {177%}
5. CEC Staff (1989)	AC	California	SO_2 + NOX + CO_2	1.69 {31%}	7.85 {95%}
6. NYPSC (1989b)	AC&RD	NY State	SO_2 + NOX + CO_2, Water, Particulates, and Land Use	0.46 {8.4%}	1.27 {15%}
7. Cavanagh (1982)	DE	Northwest U.S.	Human Health Property Damage, CO_2-Property Damage, CO_2-Famine		3.12 {38%}
8. NPPC (1989)	RD	Northwest U.S.	All but CO_2 {10% adder}	1.38 {25%}	0.83 {10%}

Table 9 (continued)
Externality Costs from New NSPS Power Plants

Average delivered cost (cents/kWh)				5.5 Natural+ Gas - CT	8.3 Coal-Steam+ Turbine
Study	Method	Location	Pollutant		
9. PSCWI	RD	Wisconsin	CCE=0.5XDC	1.04 {19%}	0.62 {7.5%}
			CCE=DC	2.07 {38%}	1.25 {15%}
10. NRDC (1989)	Other	California	CO_2	1.24 {23%}	1.25 {15%}
			All	2.62 {48%}	2.08 {25%}

+ Top figure is delivered cost in 1989 ¢/kWh. Figure below in brackets is total as % of delivered cost.

Combustion turbine (CT) is used for peaking.
Coal steam turbine is used for base load.

Methods are Direct Estimation of Costs (DE), Abatement Costs (AC), and Regulatory Determination (RD).

NSPS = New Source Performance Standards
EPRI = Electric Power Research Institute
SCAQMD = South Coast Air Quality Management District
CEC = California Energy Commission
NYPSC = New York State Public Service Commission
NPPC = Northwest Power Planning Council
PSCWI = Public Service Commission of Wisconsin
NRDC = Natural Resources Defense Council

CCE = Cost of conserved energy
DC = Delivered cost of electricity from supply resources

Source: J. Koomey, *Comparative Analysis of Monetary Estimates of External Environmental Costs Associated with Combustion of Fossil Fuels*, Lawrence Berkeley Laboratory, Berkeley: July 1990.

Using the abatement cost approach, Schilberg et al. estimated the EC for energy produced by power plants meeting new source performance standards in California, inside and outside the South Coastal Air Quality Management Districts (SCAQMDs).[11] Koomey then calculated the cost-per-unit energy with standard power plant assumptions.[12] Schilberg's values in **table 9** show that ECs for natural gas turbines inside SCAQMD were about 5 c/kWh, while in California, but outside SCAQMD, the costs were nearly 4 c/kWh. Similarly, for coal-fired steam turbines the costs were about 15 c/kWh and 8 c/kWh, respectively. Outside California, the EC for natural gas and coal-fired steam turbines were estimated to be 1.6 c/kWh and 2.7 c/kWh, respectively. These results show that the cost of abatement is greater in areas with high population density and high levels of air pollution such as the SCAQMD, which includes Los Angeles.

It should also be noted that EC estimates that exclude the effects of global warming resulting from the emission of CO_2, and from other greenhouse gases that trap solar radiation, are always much less than estimates that include CO_2 effects. This introduces a serious problem because the effects and timing of global warming are still difficult to quantify, especially at a local level. Nevertheless, global warming effects may have a profound influence on future energy policy and should not be ignored even if there are uncertainties in the quantitative effects and economic costs. Moreover, not only CO_2, but all other pollutants that contribute to the greenhouse effect, should be considered.

Summaries of ECs for existing steam power plants and for direct use of natural gas are presented in **tables 10** and **11**, respectively. Comparison of the results in these two tables with those in **table 9** shows that EC from existing power plants that do not use up-to-date pollution-control technologies are much higher than from power plants that meet current regulations. Also, as shown in **table 11**, ECs to provide thermal energy from burning natural gas directly are much lower than those for the same energy provided by electric power plants. This is an important conclusion for DSM. Electric utilities do not generally encourage customers to switch to gas because they prefer to maximize economic return on existing equipment by emphasizing peak-shaving measures, but selling all the power they can produce with base-load production levels.

Social Costs of Nuclear Power

Although the NES anticipates a substantial growth in nuclear power capacity—which currently supplies about 20 percent of all electric power for the United States—there has never been a complete assessment of the social costs of nuclear power in the United States published in the open literature. Hohmeyer has compared the social costs for wind, fossil fuel, solar photo-voltaic, and nuclear power under European conditions, but several cost factors significant to the United States—anti-terrorist measures, military requirements to guard fissionable material, R&D costs, and decommissioning of nuclear power plants—may be quite different from European conditions.[13] To quantify these costs for the United States in terms of dollars per kWh produced, it is desirable for a qualified federal agency, such as EPA or OTA, to conduct a comprehensive study.

The Hohmeyer study was conducted specifically for the Federal Republic of Germany (FRG) prior to re-unification, but the general approach should be similar for any other country with a market-oriented economy. The principal conclusion of the study was that the social costs of fossil fuel production of electric power are of the same order of magnitude as their

Table 10
Externality Costs of Electricity from
Existing U.S. Steam Power Plants in 1988

Average 1988 U.S. Electricity Price (c/kWh)				6.6	6.6	6.6
Study	Method	Location	Pollutant	Gas	Oil	Coal
1) EPRI (1987)	DE	Rural PA and WI: and 35 mi from NYC	SO_2 + NOX	0.03 {0.4%}	0.55 {8.4%}	1.11 {17%}
2) Hohmeyer (1988)	DE	West Germany	SO_2 + NOX (Low)	0.12 {1.8%}	0.37 {5.5%}	0.75 {11%}
			SO_2 + NOX (High)	0.63 {10%}	1.95 {29%}	4.02 {61%}
3) Chernick and Caverhill (1989)	AC&DE	New England	SO_2 + NOX + CO_2	2.02 {30%}	3.74 {76%}	5.93 {90%}
4) Schilberg (1989)	AC	Outside CA	SO_2 + NOX + CO_2	1.43 {22%}	2.43 {37%}	3.90 {59%}
		CA Outside SCAQMD	SO_2 + NOX + CO_2	4.68 {71%}	5.93 {90%}	11.57 {175%}
		CA Inside SCAQMD	SO_2 + NOX + CO_2	5.82 {88%}	16.00 {242%}	31.92 {483%}
5) CEC Staff (1989)	AC	California	SO_2 + NOX + CO_2	2.76 {42%}	9.13 {138%}	18.21 {275%}
6) NYPSC (1989b)	AC&RD	NY State	SO2 + NOX + CO_2	0.43 {6.5%}	0.96 {14%}	1.93 {29%}

EPRI = Electric Power Research Institute
SCAQMD = South Coast Air Quality Management District
CEC = California Energy Commission
NYPSC = New York State Public Service Commission

Source: J. Koomey, *Comparative Analysis of Monetary Estimates of External Environmental Costs Associated with Combustion of Fossil Fuels*, Lawrence Berkeley Laboratory, Berkeley: July 1990.

Table 11

Externality Costs from Direct Use of Natural Gas

Average delivered cost (cents/kWh)

Study	Method	Location	Pollutant	1.9 Average Home Gas SH & WH	1.6 Commercial Gas Boiler	1.03 Industrial Gas Boiler	1.5 Average 100% Efficiency
1. EPRI (1987)	DE	Rural PA and WI; and 35 mi. from NYC	SO$_2$ + NOX	0.0026 {0.14%}	0.0024 {0.15%}	0.0034 {0.33%}	0.0028 {0.19%}
2. Hohmeyer (1988)	DE	West Germany	SO$_2$ + NOX (Low)	0.01 {0.55%}	0.0099 {0.61%}	0.014 {1.4%}	0.011 {0.73%}
			SO$_2$ + NOX (High)	0.056 {3.0%}	0.053 {3.3%}	0.074 {7.2%}	0.061 {4.0%}
3. Chernick and Caverhill (1989)	AC & DE	New England	SO$_2$ + NOX + CO$_2$	0.49 {26%}	0.48 {30%}	0.50 {49%}	0.49 {33%}
4. Schilberg (1989)	AC	Outside CA	SO$_2$ + NOX + CO$_2$	0.33 {17%}	0.32 {20%}	0.34 {33%}	0.33 {22%}
		CA Outside SCAQMD	SO$_2$ + NOX + CO$_2$	0.61 {32%}	0.60 {38%}	0.73 {71%}	0.65 {43%}
		CA Inside SCAQMD	SO$_2$ + NOX + CO$_2$	0.72 {38%}	0.70 {44%}	0.86 {83%}	0.76 {51%}
5. CEC Staff (1989)	AC	California	SO$_2$ + NOX + CO$_2$	0.34 {18%}	0.33 {21%}	0.41 {40%}	0.36 {24%}
6. NYPSC (1989b)	AC & RD	NY State	SO$_2$ + NOX + CO$_2$, Water, particulates and land use	0.05 {2.6%}	0.048 {3.0%}	0.06 {6.0%}	0.053 {3.5%}
7. NRDC (1989)	Other	California	CO$_2$	0.17 {9.0%}	0.14 {9.0%}	0.09 {9.0%}	0.14 {9.0%}

Note: Externality costs are presented here as c/kWh. These costs, associated with natural gas, are usually given in $/MMBTU. To convert from c/kWh to $/MMBTU, multiply the number in the table by 2.93.

Source: J. Koomey, *Comparative Analysis of Monetary Estimates of External Environmental Costs Associated with Combustion of Fossil Fuels*, Lawrence Berkeley Laboratory, Berkeley: July 1990.

market costs, suggesting that the total cost of electric power is actually twice the market price paid by the consumer. For nuclear power, the social costs were twice those of electricity generated from fossil fuels. Specifically, Hohmeyer's estimates of the social costs for electricity produced from fossil fuels ranged from 2.4 c/kWh to 5.5 c/kW, while for nuclear sources they ranged from 6.1 c/kW to 13.1 c/kWh. This compares to an average market price of electricity supplied to the consumer at the time of the study in 1987 of about 6 c/kWh in the FRG.

The estimates for social costs in Europe provide an order of magnitude for energy planners in the United States until more reliable figures, based on local data, become available. The external costs in the Hohmeyer study included general economic effects, such as employment; environmental impacts, such as air pollution; governmental subsidies, such as tax deductions; and human health effects, such as respiratory diseases or radiation-induced cancers.

Nancy Rader lists some key social costs for nuclear power in the United States,[14] but does not provide an estimate in terms of dollars per unit of energy produced. The items listed include the federal subsidy under the Price-Anderson Act, which limits the nuclear industry's public liability in case of an accident. According to Bossong, the value of this subsidy is estimated to be between $1 billion and $5 billion annually in avoided insurance costs by the nuclear utilities.[15] Sandia National Laboratory estimates that a major nuclear power accident could, on average, result in monetary damages of at least $126 billion (in $1987). In addition, there would be adverse health effects due to radiation emitted in a major accident like those at Chernobyl; such effects were only narrowly averted at Three Mile Island. According to the NRC[16] and Boley,[17] the probability of a core melt down accident at one of the more than 100 U.S. reactors during the next 20 years of operation may be as high as 12 percent to 45 percent. A more complete analysis might well come up with a lower probability, but the need exists for liability insurance coverage, which is expensive.

It should be noted that uneasiness about nuclear power accidents and their social costs also have been raised recently in Japan.[18] Following the mishap at the Fukushima nuclear power complex in 1989, where a plant operator failed to shut down a 1,100 MW reactor, despite a prolonged warning signal, public trust was shaken, though no injuries were reported. In February 1991, three months after the restart of the Fukushima reactor, a more serious accident occurred at the Mihama-2 power plant near Osaka, which resulted in an abnormal leak of radioactivity, according to T. Suzuki.[19] He also cites a recent poll indicating that 90 percent of the Japanese people now feel uneasy about nuclear power and 46 percent think it is unsafe, although traditionally the Japanese public has supported nuclear power.

It is estimated by GAO that the nuclear power industry owes more than $9 billion for past uranium-enrichment services.[20] Kriesberg also claims that the nuclear industry is not paying the full cost of nuclear waste disposal,[21] which may ultimately be many times greater than the current charges and result in additional societal costs: a situation shared by the decommissioning costs for nuclear reactors. Although the nuclear industry contributes to a decommissioning fund, if this fund does not cover full decommissioning costs, taxpayers may have to pay the excess later when the reactor no longer produces electricity for the utility to sell. Kriesberg estimates that decommissioning costs for a single reactor may be as high as $1.3 billion ($1986).[22]

Tentative Conclusions

There are serious shortcomings in any attempts to quantify ECs; however, some qualitative observations can be made from available results. The ECs of burning natural gas directly to provide heat—rather than by an electrification process using coal, oil, gas, or nuclear power—are less by at least a factor of three. ECs in heavily populated areas, such as Los Angeles and areas that already have problems in meeting air emissions standards, are considerably higher than ECs in sparsely populated areas. The ECs assigned by regulatory assessments range from 3 percent to 15 percent, while the results of abatement cost analyses yield estimates that are considerably higher when expressed in terms of a percentage of resource cost. Koomey believes policies that calculate externalities based on the percentage of resource costs are inaccurate because the real damages from pollutants are not, in general, correlated with resource costs, but are strongly related to the type of pollution emission, the local topography, the population density, and other physical characteristics of the geography.[23]

It is also apparent that current estimates of social costs of electric power from combustion of fossil fuels or from nuclear fission are a substantial fraction of the price of the resource. Additional research is necessary to become more confident in the quantification of environmental externalities, particularly for nuclear power. It is desirable to develop a better understanding of the effects of externalities with respect to regional characteristics and specific resource allocations.

62

Chapter Notes

1   S.D. Cohen et al., *A Survey of State PUC Activities to Incorporate Environmental Externalities into Electric Utility Planning and Regulation* (LBL–28616, VC-224) (Berkeley: Lawrence Berkeley Laboratory, May 1990).

2   Ibid.

3   J. Koomey, *Comparative Analysis of Monetary Estimates of External Environmental Costs Associated with Combustion of Fossil Fuels* (LBL–28313) (Berkeley: Lawrence Berkeley Laboratory, July 1990).

4   J. Griffin and H. Steele, *Energy Economics and Policy*, 2nd ed. (Academic Press College Division, 1986).

5   O. Hohmeyer, *Social Costs of Energy Consumption: External Effects of Electricity Generation in the FDR* (Berlin: Springer Verlagen, 1988).

6   R. Cavanaugh et al., "A Model Electric Power and Conservation Plan for the Pacific Northwest," in *Environmental Costs* (Northwest Conservation Act Coalition, 1982).

7   J. Holdren, *Integrated Assessment for Energy-Related Environmental Standards: A Summary of Issues and Findings* (LBL-12779) (Berkeley: Lawrence Berkeley Laboratory, October 1980).

8   F. Kreith and J.F. Kreider, *Principles of Solar Engineering* (Washington, D.C.: McGraw Hill Book Co., 1980).

9   D. Goldstein et al., *Integrated Least-Cost Energy Planning in California: Preliminary Methodology and Analysis* (re: CEC docket 88–ER-8), (Natural Resources Defense Council and the Sierra Club, February 1990).

10   *Comparative Analysis of Monetary Estimates of External Environmental Costs Associated with Combustion of Fossil Fuels*, 1990.

11   G.M. Schilberg et al., *Valuing Reductions in Air Emissions into Electric Resource Planning: Theoretical and Quantitative Aspects* (re: CEC docket 88-ER-8), (JBS Energy Inc. for the Independent Energy Producers, August 1989).

12   *Comparative Analysis of Monetary Estimates of External Environmental Costs Associated with Combustion of Fossil Fuels*, 1990.

13   O. Hohmeyer, "Social Costs of Electricity Generation: Wind and Photovoltaic Versus Fossil and Nuclear," *Contemporary Policy Issues* 8, 3 (July 1990).

14   N. Rader, *Power Surge* (Washington, D.C.: Public Citizen's Critical Mass Energy Project, May 1989).

15   K. Bossong, *The Price-Anderson Act: A Multi-Billion Dollar Windfall for the Nuclear Industry* (Washington, D.C.: Public Citizen's Critical Mass Energy Project, July 1987).

16   U.S. Nuclear Regulatory Commission, NUREG/CR-2239, SAND81.

17   K. Boley, "Nuclear Power Safety Report" (Washington, D.C.: Public Citizen's Critical Mass Energy Project, December 1988).

18   T. Suzuki, "Japan's Nuclear Predicament," *Technology Review* (October 1990): 47–48.

19   Ibid.

20   U.S. General Accounting Office, "Uranium Enrichment: Congressional Action Needed to Revitalize the Program" (GAO/RCED-88-18) (October 1987).

21   J. Kriesberg, *Haste Makes Waste: Accelerated Nuclear Waste Program Could Squander Billions of Consumer Dollars* (Washington, D.C.: Public Citizen's Critical Mass Energy Project, November 1987).

22   J. Kriesberg, *Too Costly to Continue: The Economic Feasibility of a Nuclear Phase-Out* (Washington, D.C.: Public Citizen's Critical Mass Energy Project, November 1987).

23   *Comparative Analysis of Monetary Estimates of External Environmental Costs Associated with Combustion of Fossil Fuels*, 1990.

INTEGRATING
SOCIAL COSTS IN IRP

"Come on, let us reason together."

Lyndon Baines Johnson

E nergy legislation in state government is guided by recognition that the goals of economic vitality and environmental quality require efficient use of energy resources. Therefore, efforts are being made to increase energy efficiency; develop and use alternative energy resources; integrate long-term planning into a comprehensive state energy policy; and coordinate the activities of state agencies concerned with energy, economy, and environmental quality.

There are two methods for planning to meet increases in energy demand. One is to build more or larger energy-conversion systems that will use more natural fuel resources to meet the increased demand. The other is to improve the efficiency of the energy conversion system, as well as to reduce the amount of energy needed to provide given services and thereby provide consumer needs with either less fuel consumption, reduced conversion system size, or both. IRP considers these, but balances the costs of increased energy supply with the cost of reducing energy demand by improved conversion efficiency, energy conservation, or a combination of the two. To provide the necessary services at the minimum cost is not simple, but a majority of state governments require that PUCs abandon the traditional approach of considering only supply options and, instead, balance supply options with demand reduction in their overall strategy. Moreover, PUCs are asked to consider ECs or social costs when making the balance sheet between supply and demand options.

Only a few states have passed specific legislation or orders that require utilities to include the social costs of environmental and health damages resulting from energy production. Commissions in four states—Nevada, New York, Vermont, and Wisconsin—have developed numerical factors to be used by utilities in evaluating environmental externalities. Vermont and Wisconsin have assessed a cost credit for benign resources such as conservation. Vermont's formula also specifies penalties for environmental damages from certain energy sources. Nevada's PSC published a table of costs associated with emissions from energy production systems and the expected emissions of pollutants from various technologies **(tables 12 and 13)**.[1] The NYPSC gave its utilities explicit guidelines on how to treat environmental externalities when

Table 12
Costs Associated with Pollutant Emissions from Energy Production Systems in Nevada

Pollutant	Valuation (1990 dollars/lb)
Carbon Dioxide (CO_2)	0.011
Methane (CH_4)	0.11
Nitrous Oxide (N_2O)	2.07
Nitrogen Oxides (NO_2)	3.4 [1]
Sulfur Oxides (SOX)	0.78
Volatile Organic Compounds (VOC)	0.59 [2]
Carbon Monoxide (CO)	
Ambient Air Quality +	0.43 [1]
Global Warming Contribution	0.03
Total	0.46
Total Suspended Particulates/ Particulate Matter (Diameter < 10MM) TSP/PM_{10}	2.09 [1]
Hydrogen Sulfide (H_2S)	NA [3]
Water Impact	Site Specific (Determined by utility)
Land Use	Site Specific (Determined by utility)

[1]The value is applicable to EPA attainment areas. The value for an EPA non-attainment area is equal to or greater than the amount and is likely to be site-specific.

[2]The value for VOC has been adjusted to reflect the state of Nevada's status as attainment for VOC. This value is representative of an actual cost incurred in Nevada to control fugitive VOC emissions from gasoline. The value for an EPA non-attainment area is $2.75/lb.

[3]A national marginal control cost for H2S in attainment areas would be approximately $0.90/lb (OTA, 1989). The valuation of H2S is in progress at this time.

Source: Abstracted from Nevada PSC Docket, No. 89-752, 1991.

Table 13
Expected Pollutant Emissions from Various Technologies in Nevada

Electric Facilities Emissions Factors and Water Use

	Emissions (lbs/MBtu In)										Water Use (gals per MBtu In)
	NOX	SOx	TSP	CO	VOC	CO_2	CH_4	N_2O	H_2S	NH_3	
New Utility Facilities											
Baseload											
1. Combined Cycle NG	0.3933	0.0006	0.001	0.021	0.033	117	0.0019	0.0078	NA	NA	17.5
2. Combined Cycle Distillate Oil	0.5	0.315	0.001	0.018	0.0165	163	0.0016	0.0325	NA	NA	17.5
3. Coal, Pulverized, w/Scrubbers	0.6	0.6	0.03	0.024	0.004	238	0.0015	0.0325	NA	NA	48.4
4. Coal, Atmosperic Fluidized Bed	0.5	0.6	0.01	0.15	0.0028	238	0.0015	0.0325	NA	NA	1590
5. Coal, Integrated Gasification Combination Cycle	0.20	0.33	0.003	0.01	0.003	198	0.0015	0.0325	NA	NA	NA
6. MSW, Steam Boiler	0.308	0.38	0.4700	0.93	0.0300	165	0.001	0.033	NA	NA	NA
7. Wood, Steam Boiler	0.155	0.0083	0.4862	0.221	0.0773	212	0.033	0.033	NA	NA	NA
Peakers											
1. Combustion Turbine NG	0.3933	0.0006	0.0133	0.1095	0.012	119	0.012	0.018	NA	NA	0.03
2. Combustion Turbine Distillate Oil	0.6	0.212	0.03	0.116	0.0359	164	0.0016	0.0211	NA	NA	0.03
3. Reciprocating Engine, Diesel	3.3500	0.0557	0.2393	0.7286	0.2293	162	NA	NA	NA	NA	NA

Table 13 (continued)

Expected Pollutant Emissions from Various Technologies in Nevada

Electric Facilities Emissions Factors and Water Use

Emissions (lbs/MBtu Out)

	Heat Rate (Btu/kWh)	NOX	SOx	TSP	CO	VOC	CO_2	CH_4	N_2O	H_2S	NH_3	Water Use (gals per MBtu Out)
New Utility Facilities												
Baseload												
1. Combined Cycle NG	8140	3.2	0.005	0.01	0.17	0.27	952	0.015	0.063	NA	NA	142
2. Combined Cycle Distillate Oil	8140	4	2.56	0.01	0.15	0.13	1330	0.013	0.265	NA	NA	142
3. Coal, Pulverized, w/Scrubbers	9400	6	6	0.3	0.23	0.038	2240	0.014	0.306	NA	NA	455
4. Coal, Atmosperic Fluidized Bed	10000	5	6	0.1	1.5	0.03	2380	0.015	0.325	NA	NA	15900
5. Coal, Integrated Gasification Combination Cycle	9280	1.9	3.1	0.03	0.09	0.03	1840	0.014	0.302	NA	NA	NA
6. MSW, Steam Boiler	16800	5.17	6.4	7.896	16	0.504	2770	0.02	0.55	NA	NA	NA
7. Wood, Steam Boiler	16740	2.59	0.14	8.139	3.7	1.29	3550	0.55	0.55	NA	NA	NA
Peakers												
1. Combustion Turbine NG	13100	5.152	0.008	0.174	1.434	0.16	1560	0.16	0.24	NA	NA	0.4
2. Combustion Turbine Distillate Oil	13100	8	2.78	0.4	1.52	0.470	2150	0.021	0.276	NA	NA	0.4
3. Reciprocating Engine, Diesel	10000	33.500	0.557	2.393	7.286	2.293	1620	NA	NA	NA	NA	NA

Source: Abstracted from Nevada PSC docket, No. 89-752, 1991.

evaluating proposals. Weights are assigned in the scoring system to environmental costs just as to other attributes and the value of the environmental costs is then added to the avoided cost. Other states that have made efforts to quantify externalities are California, Maine, Massachusetts, and Oregon.

In addition to the methods described above, some states have adopted qualitative treatment of environmental externalities by PUCs. For example, the Arizona Public Corporation Commission (APCC) considers some environmental externalities in its LCP, but does not specify a method for quantification. Similarly, the Minnesota PUC proposed a resource planning rule that would incorporate environmental costs without giving specific instructions for quantification.

A summary of PUC activities as of April 1990 to incorporate environmental concerns into utility planning and regulation is shown in **figure 11**, based on a study by NARUC.[2] This map shows that PUCs or utilities in 17 states have adopted rules and policies to incorporate externality costs and that in eight of these states (California, Colorado, Massachusetts, New Jersey, New York, Oregon, Vermont, and Wisconsin), PUCs or utilities have developed quantitative procedures for including environmental costs in resource planning or bid evaluation.

In New Jersey, New York, and Wisconsin the adopted rules will have an immediate effect on the selection of resource options because utilities in these states will soon need to add new electric capacity. The approaches taken in some key states are summarized in **table 14** and described below.

New York. The NYPSC has prepared guidelines on treatment of environmental externalities in bid proposals to acquire new resources by the state's seven investor-owned utilities. Environmental impacts are explicitly included among the factors considered in bid evaluation. Points are assigned to environmental costs just as to any other attribute to be evaluated and the value of these environmental costs is added to the avoided costs. The approach is based on two principles key to environmental factors: all permittable projects are not environmentally equal and thus inferior projects should be penalized accordingly, and the weights for environmental factors relative to each other and relative to other non-environmental factors (price, for example) should be based on the costs to mitigate the environmental impacts. Based on a NYPSC staff analysis that assesses the environmental impact, a credit in the bidding evaluation process of up to 1.4 c/kWh (approximately 24 percent of the utility's avoided cost) is given for resources with environmental impacts lower than coal. The credits assigned to air emissions are the costs to offset or prevent those emissions at other existing facilities. For carbon dioxide, the credit is based on 20 percent of cost of reforestation to sequester the emissions. Credits for water and land use impacts are based on studies published by Bonneville Power Administration (BPA).

Figure 11
Incorporating Environmental Externalities – A State Survey

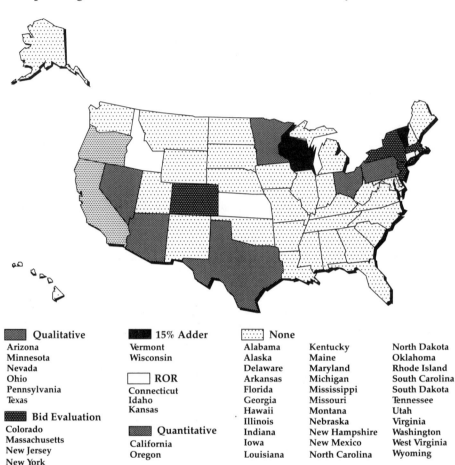

Qualitative	15% Adder	None		
Arizona	Vermont	Alabama	Kentucky	North Dakota
Minnesota	Wisconsin	Alaska	Maine	Oklahoma
Nevada		Delaware	Maryland	Rhode Island
Ohio	ROR	Arkansas	Michigan	South Carolina
Pennsylvania	Connecticut	Florida	Mississippi	South Dakota
Texas	Idaho	Georgia	Missouri	Tennessee
	Kansas	Hawaii	Montana	Utah
Bid Evaluation		Illinois	Nebraska	Virginia
Colorado	Quantitative	Indiana	New Hampshire	Washington
Massachusetts	California	Iowa	New Mexico	West Virginia
New Jersey	Oregon	Louisiana	North Carolina	Wyoming
New York				

Source: Lawrence Berkeley Laboratory, *A Survey of State PUC Activities to Incorporate Environmental Externalities into Electric Utility Planning and Regulations,* May 1990.

To arrive at a project score for the environmental component, 13 environmental attributes covering air, water, and land impacts are ranked in a scoring matrix. The distribution of weights to the 13 externalities, based on the estimated cost per kWh of mitigating them, is shown in **table 15**. The most environmentally disruptive source under the most unfavorable circumstances is assigned an environmental cost of 1.405 c/kWh. (Details of the procedure are given in a NYPSC opinion and order establishing guidelines, case 88-E-241, opinion no. 89-7, April 13, 1989.) The 1.405 c/kWh mitigating cost is on the same order as the estimated cost of energy from a new coal plant, which was established by the New York Legislature at 0.6 c/kWh.

Table 14
Status of Externalities Quantification in Selected States

State	Status of Requirements	How Requirements Are Imposed	Who Has Responsibility	Quantification Basis	Value
NY	NYPSC specified approach for ORU; similar approach likely for other	Resource bid process	Utility, based on NYPSC direction	Mitigation cost of different externalities	Up to 1.4 cents/kWh, 24 percent of avoided cost with quantity determined through scoring matrix
WI	WPSC requires utilities to give credit to non-combustion resources in planning/bidding	Least-cost planning	Utility, based on WPSC direction	Assumed benefits of noncombustion resources over combustion resources, based on NWPPC recommendation	Automatic 15 percent for noncombustion resources
VT	VPSB requires utilities to include social costs in resource option evaluations	Integrated resource assessments	Utility	Assumed benefits of DSM options	Deduct 5 percent from DSM costs for environmental benefits and 10 percent for risk reduction
MA	MADPU calls for uniform process by utilities, NEES in the lead	Resource solicitation process	Utility, with MADPU guidance	Quantifiable external impacts	NEES adds up to 15 percent to resource cost based on matrix, similar to New York approach

Table 14 (continued)
Status of Externalities Quantification in Selected States

State	Status of Requirements	How Requirements Are Imposed	Who Has Responsibility	Quantification Basis	Value
ME	MEPUC asked for comments on using societal test, including environmental benefits	Cost-effectiveness evaluation of large DSM programs	Utility	Uncertain	No values specified
OR	ORPUC requires utilities to consider externalities in cost-effectiveness	Least-cost planning	Utility	Control costs (Pacific)	Depends on utility evaluation
CA	CEC action expected	Assessment of utility resource plans	Collaborative process	Control costs (probably)	Add 10 percent to 25 percent to generation costs (proposed)

Source: L.O. Foley and A.D. Lee, "Scratching the Surface of the New Planning: A Selective Look," *The Electricity Journal* 3:6:50 (July 1990).

The New York approach is commendable, but it neglects some potential adverse effects such as those associated with the fuel extraction and transportation phases of energy production. As explained previously, some consider the mitigating cost estimates too low.

Table 15
New York's Externalities Scoring System

Externality	Mitigation Cost	Weight
Air Emissions		
Sulfur oxides	0.250	10
Nitrogen oxides	0.550	23
Carbon dioxide	0.100	4
Particulates	0.005	1
Water Impacts	0.100	4
Land Use	0.400	18
Total	**1.405**	**60**

Note: Three separate attributes are included under water impacts and six separate attributes are included under land use impacts.

Source: NYPSC, April 12, 1989, "Opinion and Order Establishing Guidelines for Bidding Program,"Case 88-E-241, Proceeding on Motion of the Commission (established in Opinion No. 88-15) as to the guidelines for bidding to meet future electric capacity needs of Orange and Rockland Utilities, Inc., 89-7; cited in LBL Report 28616.

New Jersey. Utilities in this state use an integrated resource bidding method based on an agreement between utilities' qualifying facility (QF) representatives and public utilities' staffs. The New Jersey approach uses three categories in the bidding process: economic issues (maximum of 55 percent), non-economic issues (minimum of 20 percent), and project viability (minimum of 25 percent). The non-economic issues include environmental impacts and fuel-conversion efficiency.

Wisconsin. In 1989, the Wisconsin PSC directed utilities to incorporate bidding into a LCP process to include costs associated with acid rain, global warming, human health and other factors not reflected in the utilities' LCP revenue requirement analysis (Wisconsin PSC order no. 05-EP-5, 1989). In essence, non-combustion technologies accredited for their reduced air pollution effect and non-fossil supply-side technologies and demand-side resources may cost 15 percent more than a combustion source and still be considered comparable in terms of overall social costs. While this is a step in the right direction, the order does not distinguish between various combustion resources that have different environmental impacts and it does not quantify the real cost of externalities. However, in treating life extension projects, such as refurbishing older coal-fired plants to operate beyond their originally expected life, the PSC requires utilities to include in their analysis pollution control equipment necessary to meet new source performance standards.

Vermont. The Vermont Public Service Board incorporated externalities into utilities' resource planning and acquisition processes (Vermont Public Service Board Proposal for Decision, docket no. 5270, July 1989 and docket no. 5270, April 6, 1990). The board would like to include ". . . societal costs of economic and environmental impacts which may or may not be reflected in the prices paid by utilities and participants. These costs must be considered when comparing resource alternatives." The board recognized the difficulty of proposing a clear cut formulation, but stated that, "failure to consider externalities at all is the equivalent of declaring that their significance can be quantified as equal to zero." For the present, the board requires utilities to use in their integrated assessment of supply-and-demand options a comparative risk adjustment of 10 percent and an externality adjustment of 5 percent. The 10 percent is discounted from the cost of DSM options and the 5 percent accounts for the external costs of producing the energy. The latter includes atmospheric degradation and other environmental damage. The board recognizes that this figure is not based on technical information and expects it to change when better information becomes available.

Massachusetts. The Massachusetts Department of Public Utilities requires that environmental externalities be included in evaluation of resources (Massachusetts Department of Public Utilities order no. 86-36-F, November 30, 1988). This order is an important first step, but it gives no specific direction or guidance on how externalities are to be evaluated. A later order (order no. 86-36-G, December 6, 1989) requires that utilities' evaluation of resource proposals take environmental externalities into account.

Oregon. In Oregon, the PUC places responsibility for development of environmental externality costs on the utilities. In 1989 (Oregon PUC order no. 89-507, April 1989), utilities were ordered to consider externality costs in LCP. The order required that the evaluation of resource options use both qualitative and quantitative approaches and that these costs be presented separately from conventional accounting costs to provide a range of expected tradeoffs. Recognizing that externality costs are uncertain and subjective, the process is considered flexible and open to review. Pacific Power and Light filed a long-term IRP that included environmental externalities under this order. In this plan, the company presented a scenario under which CO_2 emissions would be reduced by 20 percent from 1988 levels by 2005. In 1991, the PUC opened a new docket on externalities.

Colorado. In Colorado, environmental economic externalities are included in the QF bidding process. The Colorado PUC approved in 1988 a QF bidding process on a 100-point scale with zero to 12 points given for fuel types as shown in **table 16**. Renewables are given an additional five-point bonus at the end of the bidding process underway.

Table 16
Colorado QF Bidding—Fuel Type Credits

Fuel Type	Points
Renewables	12
Coal	5
Natural Gas	2
Oil	1

Source: "Amendment to Public Service Commission of Colorado Request for Proposals from Qualifying Facilities," Public Utility Commission of Colorado, Denver: June 9, 1989.

The Colorado PUC has adopted a new rule that provides applicants for a new line extension with a cost estimate for photovoltaic power. In performing the comparison analysis, the utility will consider line extension distance, overhead/underground construction, terrain, other variable construction costs and the probability of additions to the line extension within the life of the open extension period. Applicants must provide the utility with load data (estimated monthly kWh usage). If the ratio of estimated monthly kWh usage divided by line extension mileage is less than or equal to 1,000 (kWh/mileage < 1,000), the utility will provide the photovoltaic system cost comparison at no cost to the applicant. If the ratio is greater than 1,000, the applicant must request the photovoltaic cost estimate and may be required to pay for the analysis.

Nevada. The Nevada PUC has given specific and quantitative instruction for evaluating social costs. The commission has published a table to be used for evaluating the environmental costs associated with emission of pollutants in dollars per pound. It also has provided a table listing the anticipated emissions from various technologies in lbs/kW. These estimates, shown in **tables 12** and **13,** are the most specific guidelines for any state, but the accuracy and level of comprehensiveness have not been verified.

Chapter Notes

1   Nevada PSC, docket no. 89-752, General Session, January 1991.

2   S.D. Cohen et al., "A Survey of State PUC Activities to Incorporate Environmental Externalities into Electric Utility Planning and Regulation" (LBL-28616, VC-224) Lawrence Berkeley Laboratory, Berkeley: May 1990).

CRITERIA FOR
CONSERVATION MEASURES

"Nothing is more terrible than ignorance in action."

Johann Wolfgang von Goethe

Energy efficiency is defined as the ratio of energy required to perform a specific service to the amount of primary energy used for the process. Improving energy efficiency increases the productivity of basic energy sources by providing given services with less energy resources. For example, space conditioning, lighting, or mechanical power can be provided with less input of coal in a more energy-efficient system.

According to DOE, if energy-efficiency improvement and energy conservation in the United States were pursued vigorously and consistently with realistic energy price signals, they could contribute as much as 10 to 25 additional quads of annual energy savings by the year 2010, equivalent to 10 percent to 25 percent of the expected energy demand.[1] Achieving this potential could increase national energy security and help improve the nation's international balance of payments. Hence, improving efficiency across all sectors of the economy remains an important national objective. But DOE also notes that price signals may not be sufficient to affect energy efficiency when there are market failures. Many states are trying to decide what conservation measures should be supported to avoid such failures and are introducing significant legislation to achieve their own energy-conservation objectives, regardless of federal actions.

There is wide agreement that the cornerstone of any successful state energy strategy must be conservation of energy. However, there are many different conservation options in the energy infrastructure and they differ greatly in technical concept, ease of implementation, economic cost, and social effect. Moreover, many of these conservation measures decrease in effectiveness during use and as their scale of implementation increases. An example of the former is the decrease of electric motor efficiency as bearings wear during use and an example of the latter is using twice the thickness of insulation when an R-30 value already has been provided. It is therefore important to develop criteria to evaluate competitive conservation strategies.

Incentives or Mandates?

In framing future state energy legislation, the question arises whether incentives are preferable to mandates. Past experience may be a useful guide. In 1975, Congress passed the Energy Policy and Conservation Act (EPCA), which attempted to improve the efficiency of home appliances. The act required that energy-efficiency labels be attached to appliances to persuade purchasers to buy more energy- efficient models. Although the labels vary for different types of appliances, DOE decided that the following information must be supplied to prospective buyers:

1) Estimated average yearly cost to operate the appliance,

2) Estimated yearly cost to operate the most and least efficient similar models,

3) Estimated yearly cost to operate the appliance for different electrical rates and usage habits.

In 1989, a research program was conducted by Applied Decision Analysis, Inc. for EPA to determine the effectiveness of the energy labeling program (Energy Guide) on a national level.[2] The result of this survey indicated that "the Energy Guide labels were seen as a good idea . . . when they were first introduced, but they have not had much impact on consumer purchases." This result confirms the conclusion of other surveys conducted in the late 1970s, prior to the passage of EPCA. Consumers consistently ranked energy use and operating cost quite low on the list of attributes they consider when purchasing an appliance.

Similar observations have been made regarding the cost of utilities in buildings, especially heating and air conditioning, by prospective home buyers as well as mortgage lenders.[3] Both groups tend to emphasize the initial cost of the building—as reflected by monthly payments for principal, interest and taxes, as compared to monthly operating cost for gas and electricity—in their decision to buy or to lend. This leads to construction of energy inefficient housing and discourages installation of energy conservation retrofits in the observance of building codes or other legislation that mandates energy efficiency for buildings.

State governments have used both mandates and incentives to improve energy efficiency. Incentives are generally preferred, provided they induce the individual decisionmaker to take appropriate action. Unfortunately, in the case of buildings and appliances, the long-term economic benefits of conservation do not rank as high as the initial investment cost. Hence, to achieve increased energy efficiency, mandates may be necessary. Mandates are politically acceptable when the required actions are inexpensive, non-controversial, and simple to perform. When properly enforced, mandates have predictable results. On the other hand, to predict the effectiveness of incentives that rely on voluntary behavior, one can only use statistical experiences gained from previous similar programs; the uncertainty of the outcome is often large. But mandates and incentives can be used together. Mandates in energy conservation for buildings, for example, can be used to enforce energy-efficiency improvements such as low-flow showerheads, water heater insulation, and caulking, which are inexpensive and have rapid paybacks, while incentives can be used for improvements that are more controversial. If experience shows that incentives are not effective, state legislatures can introduce mandatory legislation. For example, electric utilities could be required to pay incentives to consumers who install energy efficient appliances or heat pumps.[4] But if the incentives offered do not lead to sufficient conservation decisions, the incentives can be increased or the conservation measure put in the form of a mandate.

Benefit-to-Cost Ratio

Concern about environmental degradation will dominate state legislative agendas in this decade. Use of any sort of energy on a large scale will inevitably cause damage to the environment (acid rain, water pollution, global warming, waste disposal, and air pollution). State legislatures are currently in the process of framing legislation designed to reduce energy consumption. But almost every energy-conservation measure requires an up-front capital investment and, given the current economic constraints in state budgets, the initial cost of an energy-conservation measure is very important. Hence, one of the criteria by which to judge an energy-conservation measure is its benefit-to-cost ratio.

There are two aspects to the benefit-to-cost ratio of energy conservation: the total value of Btus saved during the lifetime of the system to the total cost (investment, operating and maintenance) and the value of yearly net energy savings given by the difference between the energy saved and the energy used for operation and maintenance divided by the annual levelized cost of the capital investment. When the value of the saved energy and the cost of installing the conservation measure are known, the simple payback period (PP) can be calculated:

$$PP = \frac{\text{Cost of Installation in \$}}{\text{Net Value of Energy Saved per Year in (\$/yr)}}$$

The net value of the saved energy equals the amount of energy saved (Btus or kWh) times the cost per unit of the energy (\$/Btu or \$/kWh). This approach is acceptable for preliminary comparisons if the payback period is short, say less than five years. For a more precise estimate, as shown by Kreith and Kreider, the time value of money, the inflation rate, and the escalation in fuel cost must be considered.[5]

Qualitative Criteria

In addition to economic objectives, there are other criteria for which to account when evaluating energy-conservation measures for state legislation. The most important of these criteria are discussed below.

Fiscal fairness. Energy-conservation programs are usually directed at specific target audiences—for example, industry, low-income residential, high-income residential, or commercial consumers. To obtain widespread support for a conservation measure, there should be fiscal fairness; that is, if the target is the high-income residential consumer, support for the program should not be a financial burden to industry and low-income consumers. A fiscally unfair example is tax benefits for solar hot water heaters. Solar domestic hot water heaters have been installed with tax rebates primarily by high-income consumers, but the tax rebates were borne by the entire population, including the industrial sector. A fiscally fair example is a revenue-neutral "gas guzzler" tax that uses the income from the sale of fuel-inefficient cars to provide rebates to consumers who buy cars that have high mileage ratings.

Environmental impact. Nearly all conservation measures tend to improve the environment by reducing the amount of energy needed, thus reducing pollution. However, the reduction in adverse environmental impact per dollar invested varies greatly. Requiring the insulation of domestic hot water heaters, for example, will reduce the heat loss by the same amount, regardless of the energy source. But, if the hot water heater uses electricity, reducing heat losses from it will have a much more favorable environmental impact than if the water heater uses natural gas. Improving energy efficiency in the eastern states would have a greater beneficial effect on air pollution than similar improvements in a state with low population density such as Nevada or Montana.

Probability of success. Programs that have been implemented successfully under similar circumstances in other states have a higher probability of success than some others that are more innovative and are not guided by previous experience. Although innovations should be encouraged, when funding comes from public moneys and taxes, a program that has a high probability of success generally receives more support than one that is new and untried.

An extensive study of the market penetration of energy-efficiency programs was conducted in 1990 by the Oak Ridge National Laboratory.[6] The goal of this study was to refine planning assumptions and reduce uncertainties about the market penetration of DSM programs by identifying the factors that affect participation, characterizing the data required to forecast market penetration and reviewing the data available that affect program participation rates.[7] As part of the study, the authors investigated tradeoffs between incentives and marketing. The study showed that as incentives for energy conservation measures increase, the program marketing and administration costs may decrease. In some cases, paying 100 percent of the energy-efficiency measure costs reduces other program costs sufficiently to make the cost per kWh saved less than it would be at lesser incentives. In one case, the market penetration and total electricity savings resulting from a free offer to install energy-conservation blankets on electric water heaters was compared with penetration in a program with a 50 percent incentive and a shared savings offer in which the cost of the conservation measure was repaid to the utility by the customers at zero interest over a two-year period. Market penetration was five times higher for the free offer program and total cost per participant was less. Because more penetration was achieved at less cost, savings from the free offer were 10 times higher and kWh cost of saved energy was less than for the shared savings offer. Similar observations were reported for an insulation program for low-income housing in which promotional and advertising costs were greater in absolute terms than the costs for free direct installation of the measure would have been.

Ease of implementation. The ease of implementation of programs varies greatly. Hence, consideration should be given to the question of how much preparation and overhead are required to implement a given program. The number of staff required to administer the project as compared to the program's total energy saving potential is part of this question. Implementation analysis of potential energy programs would be helpful for state legislatures. Implementation analysis requires examination of political, social, and organizational constraints; including issues such as who wins and who loses, who supports or opposes the option, how agencies responded to similar proposals in the past, how the proposal will be affected by changes in economic conditions, what funding sources are available to support the program, the extent to which relevant organizations are prepared and able to accept the option, how they performed in the past, and community climate and disposition toward new ideas.

But other criteria must also be considered. For example, energy audits can easily be set up, but their success in reducing energy consumption depends on whether or not the recommendations are carried out. Experience shows that unless the audit recommendations are accompanied by funds for implementation—for example, low-income weatherization projects supported by the federal government—the energy audits alone, although easily implemented, have little positive effect on energy conservation.

The same applies to energy-efficiency ratings for appliances, windows, and other household items. Legislation that merely requires posting energy-efficiency ratings in stores or on the appliances is costly and does not have an appreciable effect on a consumer's choice to select high-efficiency items with a higher initial price, even if the differential cost could be recouped by the consumer during the life of the appliance. This can be changed in two ways: legal implementation transforming the ratings into minimum standards, and rebates that are sufficient to entice consumers to purchase the most efficient models.

Barriers to Energy Conservation

Traditional energy prices are understated because they do not include the health, social, and environmental costs of using these fuels. For example, gasoline prices do not take into account all the costs associated with military requirements to protect access to oil sources, global warming, acid rain, and adverse health effects. This is an institutional barrier to increased energy efficiency. To guide state legislators in framing effective energy-conservation legislation, some of the key barriers to increasing energy efficiency given in the draft of the NYSEP are listed below.

Lack of objective consumer information. Efficiency claims in the market place are often made by competing manufacturers, without an objective third party to evaluate the actual efficiency claims.

Failure of consumers to make optimal energy-efficiency decisions. Consumers often choose the least expensive appliance, rather than the appliance that will save them money over the long-term; consumers are also often confused about efficiency ratings and efficiency improvements.

Replacement market decisions based on availability rather than efficiency. Decisions concerning replacement of worn out or broken equipment are made without energy efficiency as a high priority. Usually the primary concern for the consumer is restoring service as quickly as possible. This requires buying whatever equipment the plumbing or heating contractor may have on hand.

Energy prices do not take into account the full environmental or societal costs. External costs associated with public health, energy production, global warming, acid rain, air pollution, energy security, or reliability of supply are usually ignored.

Competition for capital to make energy-efficiency investments. Energy-efficiency investments in the commercial and industrial sectors often must compete with other business investments; therefore, efficiency investments with a payback of more than three years are avoided.

The separation of building ownership from utility bill responsibility. Renters will rarely make energy-efficiency investments in buildings that they do not own, especially when the utilities are included in the rent.

Commercial buildings and retail space are usually built on speculation with low first-cost a priority. The building's long-term operating cost, which is usually paid by the tenant(s) rather than the owner, is not important to the speculator/builder.

Chapter Notes

1   U.S. Department of Energy, FY 1990–94 *Energy Conservation Multi-Year Plan* (Washington, D.C.: U.S. Department of Energy, 1988).

2   L.K. Carswell et al., *Environmental Labeling in the U.S.* (EPA Report ADA-89-2905) (Washington, D.C.: 1989).

3   California Energy Commission, Conservation Report No. 400–88–004, October 1988.

4   EPRI, "Compendium of Utility-Sponsored Energy Efficiency Rebate Programs," December 1989.

5   F. Kreith and J.F. Kreider, *Principles of Solar Engineering* (Washington, D.C.: McGraw-Hill Book Co., 1980).

6   L. Berry, "The Market Penetration of Energy Efficient Programs," (ORNL/CON-299) (Oak Ridge National Laboratory,Oak Ridge: April 1990).

7   Ibid.

OPPORTUNITIES
FOR STATE ACTION

"A leader has to lead, or otherwise he
has no business in politics."

Harry S Truman

I n the past decade, many state and local governments took the
leadership in the energy field and initiated legislation to deal with
energy management and conservation. Among the reasons for this
development are a desire to improve the state economy, the lack of an effective
federal energy policy, and the difficulty of formulating federal legislative action
for energy management suitable to all 50 states. There was general agreement
in testimony given during the NES hearings that the United States should have
an energy policy that stresses conservation, reduction of imported oil, and
long-term planning for research and development. Documents prepared for
the NES provide broad guidance, information and encouragement to the states
and their subdivisions to develop their own programs and funding. In the
energy policy arena, state actions are now the wave of the future.

An examination of energy-use patterns between 1970 and 1990—
published by the EIA in its annual energy review—reveals that between 1970
and 1990 (with extrapolations to the year 2000) the total energy consumption in
the building, industrial, and transportation sectors has remained nearly
constant and is not expected to increase appreciably in the next decade.
However, the electricity sector has grown from approximately 16 quads in 1970
to 25 quads in 1980 and 29 quads in 1990. It is expected that the total electricity
use in 2000 may require as much as 38 quads of primary energy.

An examination of the distribution of electricity use in the building,
industrial, and transportation sectors shows that the transportation sector used
an insignificant amount of electricity. The building and industrial sectors were
therefore the two largest users of electric power. Although the industrial sector

is a large consumer of electricity, its increase in electric power use has been relatively modest, rising from about 6 quads in 1970 to 10 quads in 1980 and remaining at that level through 1990. The industrial sector is projected to use on the order of 13 quads in 2000.

The picture looks considerably different in the building sector, however. There, the primary energy used for electricity rose from 15 quads in 1980 to 20 quads in 1990. It is projected that the building sector may use as much as 25 quads of primary energy in the form of electricity by 2000.

These figures are important for the planning of a state energy strategy because the states have essential control, through their PUCs or equivalents, over the electric utility sector—or at least a major portion of that sector. State laws and regulations can directly affect the energy efficiency of appliances and the heating requirements of buildings. Although the federal government sets some guidelines for appliance standards, states can set more stringent standards if better equipment is commercially available. States can also set standards for building efficiency that exceed federal standards. Consequently, state legislatures have the power and traditional responsibility for the portion of the energy consumption that has risen fastest.

In some ways, state legislatures have become laboratories of innovation for the federal government. Innovations that originate in state and local governments are certainly desirable, if they work. It is only through experimentation that ways and means can be identified to address the looming energy problems. However, given the limited resources of the states—and the technical complexities of energy management—many states do not have the financial capability, staff, or technical expertise to address the problem properly. Instead of analyzing the state situation on its merits, legislators often tend to see how other state legislatures address the energy problem and adopt approaches to meet their needs that have been tried in other states. No comprehensive study has been conducted to determine which of the previous approaches worked and which did not. Consequently, the evaluation of whether or not the desired outcome has been achieved is lacking and much of the legislation passed in the states today is still experimental in nature and will remain so until a method for evaluating outcomes is built into the legislative process.

There has been a good deal of debate about how much energy can be saved by improving energy efficiency. During the summer of 1989, the staffs of five national laboratories—Oak Ridge, Lawrence Berkeley, Pacific Northwest, and Idaho Engineering—made an extensive and careful analysis of the potential for energy conservation by 2010 in the residential, commercial, transportation, and industrial sectors.[1] The conclusion of this study was that an aggressive, cost-effective program could save approximately 26 percent of our national energy consumption within the next 20 years. A program that continues along current lines, and relies only on market forces, would save only 12 percent.

It should be noted that there have been some unrealistically high predictions for the potential of energy conservation to reduce electric energy use. For example, *Business Week* reported recently that Amory Lovins of the Rocky Mountain Energy Institute claims that energy conservation could reduce electric power consumption in the United States by 75 percent.[2] These highly optimistic estimates fail to mention that to achieve such orders of magnitude in savings, all existing appliances, buildings, motors, and lighting would have to be replaced by the most efficient conceivable equipment over the entire country, irrespective of cost. This kind of action would require not only astronomical amounts of money, but also an enormous amount of energy investment in equipment production and installation. If uncritically accepted by decisionmakers, such unrealistically high projections of potentially attainable energy efficiency could lead to poor plans for the future.[3]

More than 30 percent of the primary energy consumed in the United States is used in residential and commercial buildings (about 27 quads). Of that amount, the largest use for both residential (41 percent) and commercial (33 percent) buildings is space heat. It is estimated that by 2010, energy use in buildings could be as much as 39 quads. However, the use of cost-effective conservation technology could reduce this total by 20 to 30 percent, more than one-third of which would be in space heating of buildings. Because over 60 percent of all electricity generated in the United States is used by the building sector, a reduction of building heating load is of great significance to the utility sector. According to DOE's energy-conservation, multi-year plan,[4] the economically achievable conservation potential for 2010 is 2.6 quads in space heating and almost one quad in space cooling. In the commercial sector, the economically achievable conservation potential is 2.4 quads for heating and 1.2 quads for cooling. The technically achievable conservation potential for space heating is estimated to be 8 quads and 3.3 quads for space cooling.

It is important to note that the public housing sector uses over twice the energy per square foot of private-sector, multi-family buildings. Projects such as low-income housing weatherization, which have been successfully implemented by state legislation, offer future opportunities. Retrofit opportunities in the building envelope include improved insulation for walls and roofs, installation of storm windows, weatherstripping windows and doors, double-glazed windows in new buildings, replacement of single-glazed windows with double-glazed windows in existing buildings, installation of reflective insulation, nighttime temperature setbacks and many other similar conservation measures. Since the average life of buildings in the United States is 75 years, over 60 percent of today's building stock will still be in use in 2010. Emphasis in future energy savings programs should be placed on the retrofit of these buildings with cost-effective conservation technologies. DOE estimates that this could save as much as four quads annually.

88

Title III of PL 94-384, Energy Conservation and Performance Standards for New Buildings, requires that DOE issue performance standards applicable to all new buildings. These federal standards are voluntary guidelines, however, and offer states an opportunity to enforce them by other means.

Transportation Options

As can be seen from **table 17**, the mpg averages for the automobile fleets in different states vary from a low of 10.8 mpg in Wyoming to a high of 19 mpg in Hawaii. This compares to the 1987 CAFE standard of 26 mpg for a fleet of new cars. It is apparent that enormous potential exists for reducing gasoline consumption at the state level. Apart from the habits of the American public to use the private automobile extensively, policymakers face the problem of low-income consumers who generally use older and less energy-

Table 17
State Miles Per Gallon Averages, 1988

	Estimated State MPG		Estimated State MPG
Wyoming	10.8	New York	14.5
Arizona	11.8	Pennsylvania	14.5
Nevada	12.1	Montana	14.7
Iowa	12.4	Oregon	14.7
Louisiana	12.7	Alaska	15.0
South Dakota	12.9	Alabama	15.1
Tennessee	12.9	Minnesota	15.1
Mississippi	13.0	New Mexico	15.1
Kansas	13.1	Delaware	15.3
Indiana	13.3	Maryland	15.3
Rhode Island	13.4	Massachusetts	15.4
Missouri	13.5	South Carolina	15.4
Nebraska	13.5	Washington	15.4
Ohio	13.7	Colorado	15.5
Kentucky	13.8	Oklahoma	15.5
Illinois	13.9	Vermont	15.6
Florida	14.1	Virginia	15.6
North Carolina	14.1	California	15.9
Utah	14.1	Michigan	15.9
West Virginia	14.1	Arizona	16.2
New Jersey	14.2	New Hampshire	16.3
Georgia	14.3	Wisconsin	16.6
Idaho	14.4	Connecticut	16.8
Maine	14.4	Hawaii	19.0
		United States	14.2

Figures for Texas were not available.

Source: R. Lockhart, *Driving Up the Heat: A Buyer's Guide to 1991 Model Cars, Vans and Light Trucks*, 2nd ed., Washington, D.C.: Public Citizen, 1991.

efficient automobiles and spend a large fraction of their total income on gasoline. Hence, measures such as gasoline taxes may have to be offset by some kind of rebate for low-income consumers. Taxing gasoline consumption is likely to be a necessary step, however, to substantially decrease use of the single-occupant automobile and encourage the development of mass transport systems, especially in large cities. While such increased taxation may not appeal to the federal government for political reasons, a number of states have taken the initiative by taxing gasoline, proposing "fee-bate" legislation, or proposing a gas guzzler tax to encourage the use of more energy-efficient automobiles and to discourage automobile travel in general.

A 1991 survey by Public Citizen of the fuel efficiency of cars, vans, and light trucks ranked over 900 of the 1991 models by their fuel efficiency and expected lifetime gasoline consumption.[5] Government documents and trends in car and light truck efficiency data indicate great potential for future fuel savings. The study's principal findings and claims are:

1) Every four years, more oil will be saved as a result of improvements in fuel efficiency made since 1975 than is projected to be recoverable from the Arctic National Wildlife Refuge (ANWR). Setting a 45-mile per gallon standard for new cars in 2000 could further reduce gasoline consumption by at least 15 percent.

2) While fuel efficiency for new cars reached an all-time high of 28.8 mpg for model year 1988 cars, it declined to 28.1 mpg in 1990, the lowest level since 1985; the average for new domestic cars was only 26.9 mpg, while new imports averaged 29.9 mpg.

3) Out of 440 1991 model passenger cars that were also available in the 1990 model year, over 75 percent either became less fuel efficient (103 cars) or remained the same; only 109 cars achieved gains in fuel efficiency.

4) Fuel efficiency for light trucks has dropped from a high of 21.7 mpg in the 1987 model year to 21.1 mpg for 1991 model trucks.

5) Large cars are not necessarily the least fuel efficient nor are smaller cars always the best; there is a wide variation in fuel efficiency within each size class. In fact, the average fuel efficiency of new cars on the road could have been 33.4 mpg in 1991—nearly 20 percent higher than the average of 28.2 mpg for all 1991 cars—if consumers chose only the most efficient models in each size class.

6) Data from the past 10 years indicate that more fuel-efficient cars can be as safe as larger and less fuel-efficient cars. For example, research safety vehicles, produced for the U.S. Department of Transportation using 1975 technologies, could achieve a fuel efficiency of 43 mpg and high levels of passenger protection in 40 mph crashes—better than many production vehicles on the road today.

7) Certain available cost-effective technologies, such as aerodynamic improvements, four valves-per-cylinder engines, front-wheel drive, multi-point fuel injection, electronic transmission controls, and intake valve controls, can dramatically improve fuel efficiency without vehicle size reductions.

8) The five states that use the most gasoline to power motor vehicles are California, Florida, New York, Ohio, and Texas. The five states that consume the largest amount of gasoline per capita are Georgia, Montana, North Dakota, South Dakota, and Wyoming.

Appendix 1 presents quantitative gasoline-consumption data, per state and per capita, for each state.

Appendix 2 presents quantitative vehicle-miles-travelled (VMT) data, per state and per capita, for each state.

It is widely believed that energy taxes have the potential to be effective policy tools for energy management. Such taxes encourage conservation and provide funds needed to support environmentally responsible energy sources. Sweden taxes fossil fuel-produced electricity at 300 percent of the cost of producing that electricity—in an attempt to monetize the true social costs of producing electricity. Substantial gasoline taxes have been in place in Europe and Japan for some time. The taxes range from $1.44 to $3.56 per gallon, compared with combined federal and state tax averages in the United States of around 35 cents per gallon.[6]

Another potential source of energy in the transportation area is oil reuse. Technology exists to recondition used automotive oil, but it can also be used as fuel in some industrial operations. In 1989, EPA published a guide for establishing local recycling programs, but left implementation and legislation to the states. **Table 18** shows what actions states have taken to collect and reuse oil. In California, 46 percent of all used oil—approximately 63 million gallons—was recycled in 1988, while a volunteer program in Virginia reclaimed 340,000 gallons the same year.

Appliances and Lighting

Another area for energy saving is lighting. Changing from traditional incandescent lights to more efficient fluorescent lights can save energy and money in homes, businesses, industries, and state buildings. Replacement of inefficient mercury vapor street lights with sodium vapor lights has been a successful investment for the city of Decorah, Iowa. The city replaced 319 lights at a total cost of $71,000, resulting in a yearly savings of 160,000 kWh valued at more than $19,000. The energy cost savings from the new lights will pay for them in 3.7 years.

Table 18
Used-Oil Collection Programs and Legislation in 15 Selected States

	Tax or deposit on retail oil	Retail sign rule (1)	Partial liability protection for collectors	Landfill disposal banned (2)	Program operator, funding	Estimated collection locations	Comments	Contact person
Alabama	No	No	No	No	Project ROSE (University of (Alabama); state funding	175	Established 1977; one of first programs to focus on education	Janet Graham (205) 348-1735
California	No	Yes	No	Yes	State operates, funds	700–1,000	Used oil considered hazardous waste	Herb Berton (916) 322-2651
Florida	No	No	Yes	Yes	State, federal oil-overcharge funds	400	Several innovative local programs, including curbside and mobile collection	Betsy Galocy (904) 488-0300
Georgia	No	No	No	Yes	Project PETRO (state), federal oil-overcharge funds	175	Education program emphasizes toll-free phone number for collection locations	Mark Mixon (404) 656-2833
Iowa	No	Yes	No	Yes	No formal program	75	State requires 8% insurance rate discount for underground storage tanks when used oil accepted from public	Robert Craggs (515) 281-8408

Table 18 (continued)
Used-Oil Collection Programs and Legislation in 15 Selected States

	Tax or deposit on retail oil	Retail sign rule (1)	Partial liability protection for collectors	Landfill disposal banned (2)	Program operator, funding	Estimated collection locations	Comments	Contact person
Louisiana	No	No	No	Yes	No formal program	25	Some large companies provide collection sites for employees' DIY used oil	Tom Patterson (504) 342-9081
Maryland	No	Yes	Yes	Yes	State operates, funds	320	State provides collection tanks for 120 municipal sites	Pat Tantum (301) 974-7254
Massachusetts	No	(3)	No	Yes	State operates, funds	1.030	State starting to gear up education program; used oil listed as hazardous waste	Cynthia Bellamy (617) 292-5848
Michigan	No	No	No	Yes	West Michigan Environmental Action Council, state funding	700	State used oil policy being revised, under legislative mandate	Julie Stoneman (616) 451-3051
Minnesota	No	Yes	No	Yes	State operates, funds	1,500	State gives counties up to $5,000 for collection tanks	Randy Hukriede (612) 643-3470
Oregon	No	Yes	No	No	No formal program	200–300	In addition, 100 cities have curbside programs	Peter Spendelow (503) 229-5253

Table 18 (continued)
Used-Oil Collection Programs and Legislation in 15 Selected States

	Tax or deposit on retail oil	Retail sign rule (1)	Partial liability protection for collectors	Landfill disposal banned (2)	Program operator, funding	Estimated collection locations	Comments	Contact person
Rhode Island	Yes; 5¢/ quart tax	No	Yes	Yes	State operates, funds; also, federal oil-over-charge funds	29	State provides special containers to local governments; used oil listed as hazardous waste	Eugene Pepper (401) 277-3434
Vermont	No	No	Yes	Yes	State operates, funds; also, federal oil-over-charge funds	40	Used oil collected in state program must be re-refined, not burned for fuel; state subsidizes	Barb Winters (802) 244-7831
Washington	No	Yes	No	No	State operates, funds	250	Education program includes toll-free phone number for collection locations	Steve Barrett (206) 459-6286
Wisconsin	No	Yes	No	Yes (4)	State operates, funds	150–200	Counties responsible for having collection sites; state offers technical assistance only	Paul Koziar (608) 266-5741

(1) Regulations requiring oil retailers to post signs with recycling information or collection locations.
(2) In most cases, road oiling is also banned; Iowa is one exception.
(3) All Massachusetts oil retailers must accept used oil originally purchased in their store; receipt required.
(4) Effective January 1, 1991.

Source: *Resource Recycling,* 1990.

One of the areas targeted by many state legislators is the residential appliances market. The results of an analysis of the conservation potential presented by OTA's Peter Blair are shown in **figure 12** for refrigerators, central air conditioning, water heaters, and electric ranges.[7] In this graph the yearly energy consumption for the average and the best current models, as well as the potential of technologically advanced units, are depicted. For example, the

Figure 12
U.S. Residential Appliances Energy-efficiency Potential

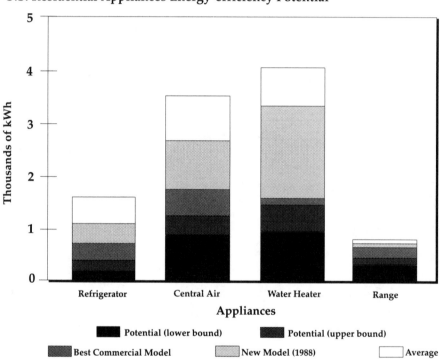

Source: P. Blair, *Energy Efficiency and Electric Use in the 1990s*, OTA, March 1990.

energy used by the average refrigerator is 1,500 kWh per year, while the best commercially available model would consume only about 700 kWh—clearly an enormous conservation potential subject to state legislation.

Many states, as well as the federal government, have proposed that standards for household appliances be developed and that the performance of appliances be either posted or attached to the item at the time of sale. It is questionable, however, whether consumers respond to posted technical performance on the appliance by buying the more energy-efficient model, rather than the cheaper model. There is evidence that it is more effective to require certain minimum appliance performance standards for refrigerators, ovens, toasters, and freezers to qualify for sale in the state.

General Observations

According to a study released by Public Citizen, in 1990 the five states that used energy most frugally were Arizona, California, Colorado, New York, and Vermont.[8] The average yearly energy consumption per capita in these states was 239 MBtu. At the other end of the spectrum were Alaska, Kansas, Louisiana, Texas, and Wyoming. If every state consumed the same amount of energy per capita as the five most energy-efficient states, energy use in the United States would be 35 percent less. If every state consumed petroleum at the same rate per capita as do the five lowest consumer states, petroleum consumption in the country would be 27 percent less. The average per capita energy consumption in 1990 was 323 MBtu per person. This is equivalent to approximately 55 barrels of oil per capita each year. If the 23 states that use more energy than the national average would reduce their consumption to the average, nationwide consumption would be nearly 12 percent less.

These data show that there is a great potential in many states to increase energy use efficiency appreciably with available methods. According to energy plans for some of the most energy-efficient states, energy consumption can be further reduced. OTA has compiled a list of energy-saving technologies that could significantly improve U.S. energy efficiency. **Table 19** provides a partial list of some of the technologies that are already commercially available and could be implemented. Some of these options have been discussed in more detail elsewhere in this study. Further information can be obtained in the references cited as footnotes in **table 19**.

From a policy perspective, it is important to recognize that the energy savings offered by new technologies are uncertain. One must consider the cost, changes in lifestyle, relation to other public goals, and investment in energy required to construct and install the energy-saving technologies. For instance, switching from standard to more energy-efficient ballasts in lamps will require initial budgetary outlay to achieve long-term savings. Staging the replacement will help, but it must be carefully orchestrated.

Table 19
Commercially Available Technologies that Improve Energy Efficiency

Residential/Commercial

- Switching from standard fluorescent ballasts to more efficient solid-state electromagnetic ballasts in lighting decreases energy use by 20 to 25 percent; adding an optical reflector to fluorescent lamps increases useful light output by 75 to 100 percent, cutting energy use by 30 to 50 percent.[1]
- It is possible to develop windows with thermal insulation equivalent to three inches of fiberglass.[2]

Table 19 (continued)
Commercially Available Technologies that Improve Energy Efficiency
- The efficiency of most home appliances (refrigerators, freezers, central air conditioners, electric water heaters) can be nearly doubled by using technology already on the market.[3]
- Demonstration homes in Minnesota that use new insulation techniques use 68 percent less heat than the average U.S. home.[4]
- Installing Variable Air Volume (VAV) systems that react to changes in heating and cooling needs by adjusting the amount of air-conditioning can generate savings from 25 to 80 percent over standard systems.[5]
- Information technologies, such as Energy Management Systems (EMS), can be applied to optimize the heating and cooling needs of a building. These systems range from simple timers to sophisticated microprocessor based systems. Computerized EMS typically provide a 10 to 20 percent savings.[6]

Automobiles[7]

Available New Technology	Prevailing Technology	Fuel Savings Technology (percent gain)
4 valves/cylinder	2 valves/cylinder	10
Turbocharging	standard carburetor	5–10
Fuel injection	standard carburetor	6
Continuously variable transmission	3-speed automatic	10
Overdrive	3-speed automatic	7
Aerodynamic design	15% reduction in drag	3

Industry

- Electrically driven freeze process is estimated to use one-eighth as much energy as the fuel-based evaporators.[8]
- Recovery of waste heat in the chemical industry has reduced energy use per pound of product by 43 percent since 1974, and the potential for further cuts of 32 to 48 percent exists.[9]
- The use of ultraviolet radiation to dry paint and cure plastic resins reduces curing time from 20 minutes to 1/15th of a second.[10]
- Use of continuous casting technology, as opposed to ingot casting, in the steel industry reduces energy consumption by half and increases product yield from 80 to 95 percent.[11]
- Adjustable-speed drives already in application get energy savings of 20 to 25 percent in compressors, 30 to 35 percent in blowers and fans and 20 to 25 percent in pumps.[12] It is estimated that on average, adjustable-speed motors cut electricity requirements by one-fifth.[13]

Source: Office of Technology Assessment, Background Paper: "Energy Use and the U.S. Economy," June 1990.
Note: See table 19 notes after the chapter notes.

Specific Opportunities for Legislation

There are excellent opportunities in almost every state to initiate measures that can reduce energy consumption. Economic implications will vary from place to place because predictions of future energy demand and cost are uncertain. But there are energy conservation strategies that have been successful in the past and this report has documented many of them. The following list of ideas for such strategies is not necessarily suitable for every state. The list is, rather, a menu of ideas and should be carefully scrutinized before implementing any one of them. But they can be a starting point for further deliberation and subsequent action.

Transportation sector

- Promote construction of mass transit systems
- Provide initiatives for ride sharing
- Implement installation and use of intelligent vehicle/highway systems
- Coordinate traffic light sequencing with traffic flow by use of responsive computer programs
- Support expansion of telecommuting
- Provide funds for scrapping inefficient and polluting vehicles
- Purchase alternative fuel vehicles for state agencies
- Introduce "gas guzzler taxes"
- Increase gasoline taxes
- Establish information sharing on alternative fuels with other states

Building sector

- Require energy-efficient mortgage plans for home buyers
- Mandate improved energy efficiency in state buildings
- Implement model building standards and inspection for new buildings
- Improve public understanding of the economics of energy efficiency
- Provide technology transfer of new building technologies
- Mandate appliance efficiency standards
- Legislate use of high-efficiency lighting systems
- Provide home energy audits with low-cost financing for recommended improvements

Utilities sector

- Mandate integrated resource planning
- Mandate use of least-cost bidding systems by PUCs
- Encourage reasonable profits for utilities that implement demand-side reduction

- Encourage small-scale hydrosystems and other cost-effective renewable resources
- Incorporate environmental externalities and social costs in utility bidding process
- Facilitate siting of municipal solid waste-to-energy incineration systems
- Encourage regional electric power planning and transmission
- Provide tax incentives for energy conservation measures

Industrial sector

- Subsidize industrial energy audits with low-cost loans for implementation of recommended procedures
- Mandate use of high-efficiency lighting and motors
- Provide incentives for source reduction of imported and toxic materials
- Provide for environmentally safe incineration of industrial wastes for energy
- Provide low-interest loans for development of plants that use recycled materials (such as de-inking of paper)
- Encourage cost- and energy-effective recycling
- Develop markets for recycled materials (such as procurement incentives by state agencies)

General recommendations

- Improve math and science education
- Require scientific and technological literacy of state employees
- Develop videos and television programs to educate public about energy

These and other ideas for increasing energy-use efficiency will be considered by state legislatures in the forthcoming legislative sessions. Evaluation of the effectiveness of specific conservation measures under given conditions will require technical expertise that is not always available, especially in smaller states. Information sharing between states and federal support for quantitative evaluation of the expected saving from various measures under consideration by state legislatures would help state governments in the development of good energy legislation.

Chapter Notes

1   R.S. Carlsmith et al., "Energy Efficiency: How Far Can We Go?" (ORNL/TM-11441) (Oak Ridge: Oak Ridge National Laboratory, 1990).

2   E. Smith, "Amory Lovins' Energy Ideas Don't Sound So Dim Anymore," *Business Week* (September 16, 1991): 92.

3   A.D. Rossin, "Letter to the Editor," *Science* 251 (March 15, 1991): 1296.

4   U.S. Department of Energy, *FY 1990–94 Energy Conservation Multi-Year Plan* (Washington, D.C.: 1988).

5   R. Lockhart, "Driving Up the Heat: A Buyer's Guide to 1991 Model Cars, Vans and Light Trucks," 2nd ed. (Washington, D.C.: Public Citizen, 1991).

6   C. Flavin and N. Lenssen, "Designing a Solar Economy: A Policy Agenda," *Solar Today* (May/June 1991): 18–20.

7   P. Blair, "Energy Efficiency and Electricity Use in the 1990s" (Washington, D.C.: Office of Technology Assessment, March 1990).

8   J. Becker et al., *Energy Audit: A State-By-State Profile of Energy Conservation and Alternatives* (Washington, D.C.: Public Citizen's Critical Mass Energy Project, October 4, 1990).

Table 19 Notes

1   H.S. Geller, "Commercial Building Equipment Efficiency: A State-of-the-Art Review," for the Office of Technology Assessment (May 1988): 12; Electric Power Research Institute, "Lighting the Commercial World," *EPRI Journal* (December 1989):12–13; and J.H. Gibbons et al., "Strategies for Energy Use," *Scientific American* (September 1989):140.

2   S. Selkowitz, "Window Performance and Building Energy Use: Some Technical Options for Increasing Energy Efficiency," in *Energy Source: Conservation and Renewables*, D. Hafemeister, H. Kelly, and B. Levi (eds.) (New York: American Institute of Physics, 1985).

3   H. S. Geller, "Residential Equipment Efficiency: A State-of-the-Art Review," for the Office of Technology Assessment (December 1987): 3.

4   J.H. Gibbons et al., 141.

5   H.S. Geller, "Commercial Building Equipment Efficiency," 10.

6   Ibid., 8.

7   K.G. Duleep, Energy and Environmental Analysis, Arlington, Virginia, "Developments in the Fuel Economy of Light-Duty Highway Vehicles," for the Office of Technology Assessment, August 1988.

8   M. Ross, "Improving the Energy Efficiency of Electricity Use in Manufacturing," *Science* (April 21, 1989): 244.

9   U.S. Department of Energy, Oak Ridge: Oak Ridge National Laboratory, "Energy R&D: What Could Make a Difference?" 2, part 1 (May 1989): 71.

10   C.A. Berg, "The Use of Electric Power and the Growth of Productivity: One Engineer's View," draft, Northeastern University, Boston, 33.

11   "Energy R&D: What Could Make a Difference?" 86.

12   S.F. Baldwin, "The Materials Revolution and Energy-Efficient Electrical Drive Systems," *Annual Review of Energy*, 13 (1988): 87.

13   A. Kahane and R. Squitieri, "Electricity Use in Manufacturing," *Annual Review of Energy*, 12 (1987): 236.

POSTSCRIPT

"Energy efficiency can only buy time for the world to develop 'low energy paths' based on renewable sources which should form the foundation of the global energy structure during the 21st century."

Harlem Brundtland, Prime Minister of Norway

The energy-management and conservation measures outlined in this book can reduce the demand for energy. But energy efficiency and conservation will not solve the looming energy crisis in the long run. In the short term, they will bring about real economic benefits and help preserve the limited fossil fuels for alternative uses, such as chemical feed stocks; however, increased energy efficiency and conservation cannot generate energy or provide new sources of power. There are only two major energy sources that can be considered technologically viable and ecologically acceptable for the longer term—renewable solar and nuclear. Renewable solar sources include biomass, solar thermal, photovoltaics, wind power, ocean energy, and hydroelectric power. These sources, combined with nuclear energy, represent the only long-term options for a safe and secure energy supply. Fusion may some day be developed as a virtually limitless power source, but to date it is only a concept for research, not a viable technology.

The end of the fossil fuel age is in sight and we must redirect the world's energy economy in the next few years to ensure a sustainable future. Technologies to initiate this transition have been demonstrated all over the world; except for wood and waste incineration, hydroelectric, wind, and nuclear power, however, they have not achieved significant market penetration. The reason is that solar renewable technologies face political, economic and institutional barriers. Rules governing today's energy economy were developed during a time when the goal was to increase energy use and expand fossil fuel consumption. To change these rules will require significant policy changes that must include reducing subsidies for fossil fuels; internalizing social costs in energy prices; imposing taxes on fossil fuels that reflect security, environmental and replacement costs; increasing support for research and development of efficient and renewable technologies; providing reliable information on the tradeoffs and real costs of different energy options; educating the public about the need for solar and nuclear power; and strengthening federal and state policies to facilitate the necessary transition. These are enormous tasks that require funding, leadership, and direction from the federal government, and worldwide cooperation. Moreover, the future supply of energy and the quality of this environment are intimately linked to the world's population.

State and local governments can play an important role, but they alone cannot reorder our basic energy priorities. State governments are best equipped to promote energy savings in housing, transportation, land use planning, and solid waste management. The capital and energy investments required for a basic change in our energy infrastructure are immense. Individual states that are already financially strapped will not be able to provide such funds. The only source of funding sufficient for this task at present seems to be the hoped-for savings from a reduction in military spending. But a redirection of the expected "peace dividend" toward an efficient and renewable energy infrastructure can come only from the federal government.

For the government to take such drastic action requires a clear indication that a majority of the public is willing to support it. Achieving a sustainable and efficient energy future will require the leadership of the federal government, with cooperation from state governments and the support of consumers and industry. It will be painful to choose between short-term and long-term objectives. But to achieve a viable energy infrastructure, we must set realistic long-term objectives that take into account the nation's energy sources, available means to achieve chosen goals, and the limits nature imposes on man and his environment. We must strive to achieve national consensus and establish a comprehensive energy strategy that can shape a sustainable energy future for all.

BACKGROUND FOR CASE STUDIES

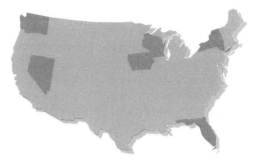

Increased energy-use efficiency reduces air emissions, improves the productivity of energy use, enhances economic competitiveness, reduces the need for imported foreign oil at a time when domestic production is declining, reduces the nation's trade deficit, reduces electric utility load growth and the need for new nuclear or fossil-fueled plants, and serves the consumer by providing the same—if not improved—services at a lower cost.[1] Energy-efficiency policies should be considered an integral part of economic development. Energy markets have changed in recent years, and most states realize that energy efficiency and economic development are directly related. For most of this century, state economies have thrived and depended on low-cost petroleum for the majority of their energy needs. Low energy costs make many energy-efficiency programs less competitive, but when oil prices are higher ($30-$50 a barrel), more energy-efficiency programs are competitive.

It is no surprise that states known for advances in energy efficiency import energy from sources outside the state. With billions of dollars available for existing local economies to import energy supplies, it makes sense to produce as much energy as possible within the state, while minimizing the energy dollars paid to sources outside the state. For example, one-half of New York's 1989 $28 billion annual energy expenditures went to out-of-state sources.[2]

Energy efficiency is increasingly important to state businesses. Energy cost is one of the most important factors to influence decisions about where to locate new enterprises—often ahead of site characteristics, business climate, and taxes. Improvements in energy efficiency can help state industries that face short-term economic difficulties become more productive and competitive. By lowering operating costs, high energy use efficiency also improves the financial bottom line of businesses. Providing a reliable, adequate, and efficient supply of energy for a state's businesses and citizens is sound fiscal policy.

Numerous states have been active on the conservation/energy efficiency front for decades. State lawmakers have promoted energy efficiency for years through a number of strategies, including state procurement standards that evaluate energy costs as part of life-cycle costing methods, building energy codes, tax credits for efficiency measures, tax liability for producing fossil fuels (severance taxes), residential building codes, mass transit incentives, appliance-efficiency standards, provision of funds for research and development of energy-efficient products, and control of state utility commission rates charged for electricity.

In recent years, the availability of oil-overcharge funds have made energy-efficiency programs possible. These funds represent fines paid by oil companies found to have violated federal price regulations in the late 1970s. A portion of the fines are paid to the U.S. Department of Energy (DOE) and dispensed to the states—which are charged with providing restitution, in the form of energy-related programs—to overcharged citizens. The proportion of oil-overcharge funds a state receives is usually based on petroleum consumption patterns at the time of the price violations.

These funds are drying up and other primary funding sources for state energy programs—private foundations, state general funds, and separate federal grants—must be found to fill the void. Federal support is never certain, and 30 states expect revenue shortfalls in fiscal year (FY) 1992. State legislators want to know how to finance future programs and which energy programs are the most cost-effective, so that limited available dollars go to the most appropriate programs.

Organization Of The Case Studies

This part of the book addresses three key areas in the energy efficiency field. It presents the advances of six states (Florida, Iowa, Nevada, New York, Washington, and Wisconsin) in integrated resource planning (IRP) and energy conservation. It lists and summarizes the most important energy-efficiency legislation from each of these six states. It evaluates state energy programs—that sometimes resulted from state legislation—using a matrix with both quantitative and qualitative criteria. Each state is discussed separately under five sections.

State Background. Section one discusses general energy consumption trends and provides background energy information on the state. A graph that shows the state's total energy consumption pattern from 1970 to 1988, and a bar chart that shows the state's energy use by source (coal, natural gas, petroleum, nuclear, hydroelectric, imported electricity) is included in this section. The per capita energy consumption from 1970 to 1988 for all six states is plotted in **figure 13**.

Figure 13
Per Capita Energy Consumption by State, 1970–1988

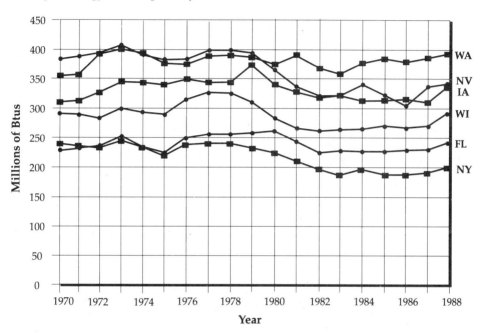

Source: Energy Information Administration, *State Energy Data Report, Consumption Estimates, 1960–1988,* DOE/EIA-0219, 1988.

Energy Consumption By Sector. Section two addresses more specific energy consumption patterns in the four traditional sectors: industrial, transportation, residential, and commercial. In the interest of uniformity and objectivity, NCSL uses energy-consumption statistics compiled by the Energy Information Administration (EIA). The Washington and Wisconsin energy offices provided NCSL with energy consumption statistics that varied from the EIA information. These differences are noted in the text. **Figure 14** shows 1988 per capita energy consumption by state, and total energy consumption by state is shown in **figure 15** for all bar charts. **Table 20** shows energy consumption by sector and consumption per capita for all states in 1988.

Table 20
Energy Consumption by Sector and Consumption per Capita, Ranked by State, 1988

(Sectors: Trillion Btu; Consumption per Capita: Million Btu)

Rank	Residential Sector		Commercial Sector		Industrial Sector		Transportation Sector		Consumption per Capita	
1	California	1,254.7	California	1,206.5	Texas	5,324.2	California	2,711.9	Alaska	991.3
2	Texas	1,109.4	New York	994.9	Louisiana	2,212.4	Texas	2,156.4	Wyoming	786.2
3	New York	1,024.6	Texas	992.8	California	1,796.8	Florida	1,110.4	Louisiana	782.8
4	Illinois	881.7	Illinois	664.4	Ohio	1,534.7	New York	874.0	Texas	569.3
5	Ohio	857.9	Florida	835.6	Pennsylvania	1,384.5	Pennsylvania	859.7	North Dakota	466.6
6	Pennsylvania	847.9	Ohio	581.9	Illinois	1,255.4	Ohio	810.3	Indiana	445.7
7	Florida	750.5	Pennsylvania	509.1	Indiana	1,197.9	New Jersey	786.3	Kansas	423.4
8	Michigan	718.9	New Jersey	451.5	Michigan	910.6	Illinois	775.5	Montana	414.9
9	New Jersey	524.2	Michigan	435.5	Alabama	756.2	Louisiana	704.5	West Virginia	414.6
10	North Carolina	457.1	Virginia	360.8	Tennessee	697.3	Georgia	692.4	Oklahoma	395.7
11	Indiana	449.5	North Carolina	319.7	New York	692.1	Michigan	688.3	Alabama	393.4
12	Georgia	434.2	Massachusetts	315.1	Washington	640.3	Virginia	590.1	Washington	388.5
13	Virginia	431.2	Georgia	303.4	Georgia	608.2	North Carolina	576.2	Kentucky	376.0
14	Massachusetts	399.0	Washington	294.1	North Carolina	593.5	Indiana	558.0	Mississippi	358.1
15	Missouri	394.5	Missouri	287.0	Kentucky	588.5	Washington	513.4	Idaho	353.6
16	Tennessee	376.3	Indiana	272.4	New Jersey	524.1	Missouri	475.2	Tennessee	350.0
17	Wisconsin	363.6	Louisiana	241.7	Oklahoma	510.8	Tennessee	461.1	Delaware	348.5
18	Washington	359.3	Wisconsin	229.8	South Carolina	486.2	Massachusetts	417.6	Ohio	348.3
19	Maryland	325.2	Colorado	226.1	Virginia	458.1	Kentucky	381.0	New Mexico	347.0
20	Minnesota	321.2	Arizona	199.3	Wisconsin	457.3	Alabama	373.7	Nevada	337.1
21	Alabama	295.0	Oklahoma	191.7	Minnesota	455.0	Minnesota	347.2	Nebraska	334.2
22	Louisiana	291.2	Minnesota	191.6	Florida	432.3	Maryland	343.6	Iowa	332.6
23	Kentucky	277.3	Alabama	189.3	West Virginia	412.7	Wisconsin	341.2	South Carolina	329.9
24	Oklahoma	243.6	Tennessee	179.5	Kansas	412.7	Oklahoma	333.6	Arkansas	329.8
25	Connecticut	235.8	Kentucky	174.1	Maryland	390.5	Arizona	327.6	Georgia	321.5
26	South Carolina	230.7	Maryland	173.0	Missouri	354.4	Mississippi	319.4	Oregon	317.7
27	Iowa	222.9	Connecticut	168.5	Mississippi	354.3	Kansas	293.8	Utah	315.5
28	Colorado	205.8	Kansas	164.9	Iowa	347.9	Oregon	280.3	Maine	308.6
29	Arizona	196.3	South Carolina	157.1	Arkansas	297.2	Colorado	273.2	Illinois	306.0
30	Oregon	192.8	Oregon	156.1	Alaska	276.4	South Carolina	269.2	Virginia	306.0
31	Kansas	185.3	Iowa	138.8	Oregon	250.4	Iowa	232.2	Minnesota	305.2
32	Mississippi	185.8	Nebraska	113.5	Wyoming	216.0	Arkansas	225.7	Pennsylvania	300.2

Table 20 (continued)
Energy Consumption by Sector and Consumption per Capita, Ranked by State, 1988

(Sectors: Trillion Btu; Consumption per Capita: Million Btu)

Rank	Residential Sector		Commercial Sector		Industrial Sector		Transportation Sector		Consumption per Capita	
33	Arkansas	164.1	Arkansas	103.2	Massachusetts	214.5	Connecticut	216.4	North Carolina	300.0
34	West Virginia	131.2	New Mexico	100.4	Utah	200.9	New Mexico	193.8	Michigan	298.0
35	Nebraska	127.0	Mississippi	96.5	Colorado	194.9	Nebraska	160.2	New Jersey	296.2
36	Utah	96.4	West Virginia	85.5	Arizona	174.8	Alaska	152.2	Missouri	294.0
37	Maine	84.8	Utah	84.5	New Mexico	157.2	Utah	151.8	Wisconsin	288.1
38	Nevada	74.7	District of Columbia	75.4	North Dakota	152.4	West Virginia	148.3	South Dakota	284.4
39	Idaho	73.9	Idaho	69.0	Montana	141.6	Hawaii	143.5	District of Columbia	283.1
40	New Mexico	72.8	Nevada	63.3	Connecticut	136.0	Nevada	133.2	Colorado	273.3
41	New Hampshire	72.5	Maine	53.2	Nebraska	134.9	Maine	121.7	Maryland	268.1
42	Rhode Island	58.2	Alaska	51.6	Idaho	126.5	Idaho	85.6	Arizona	257.8
43	Montana	57.9	Montana	50.3	Maine	112.6	Montana	84.2	California	246.1
44	South Dakota	54.2	Rhode Island	44.6	Delaware	86.0	Wyoming	84.1	Hawaii	242.5
45	North Dakota	54.2	Wyoming	41.3	Nevada	84.1	New Hampshire	77.4	Florida	237.4
46	Delaware	47.0	New Hampshire	38.8	Hawaii	82.3	North Dakota	68.6	Connecticut	234.2
47	Alaska	40.1	Hawaii	36.1	New Hampshire	54.1	South Dakota	64.5	Vermont	231.5
48	Wyoming	36.1	North Dakota	36.0	South Dakota	52.4	Delaware	62.9	Massachusetts	228.6
49	District of Columbia	38.0	Delaware	32.1	Rhode Island	43.8	Rhode Island	61.7	New Hampshire	223.8
50	Vermont	35.0	South Dakota	32.0	District of Columbia	31.5	Vermont	44.3	Rhode Island	217.9
51	Hawaii	24.0	Vermont	23.8	Vermont	26.0	District of Columbia	30.3	New York	200.2
	United States	16,371.0	United States	12,641.7	United States	28,045.4	United States	22,187.3	United States	326.5

Source: EIA, *State Energy Data System*, 1988.

Figure 14

Per Capita Energy Consumption by State, 1988

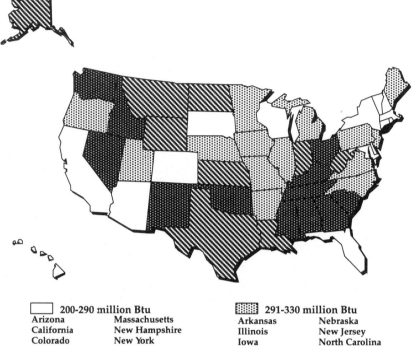

□ 200-290 million Btu		▒ 291-330 million Btu	
Arizona	Massachusetts	Arkansas	Nebraska
California	New Hampshire	Illinois	New Jersey
Colorado	New York	Iowa	North Carolina
Connecticut	Rhode Island	Maine	Oregon
Delaware	South Dakota	Michigan	Pennsylvania
Florida	Vermont	Minnesota	Utah
Hawaii	Wisconsin	Missouri	Virginia
Maryland			

▓ 331-400 million Btu		▨ 401-1,000 million Btu	
Alabama	New Mexico	Alaska	North Dakota
Georgia	Ohio	Indiana	Texas
Idaho	Oklahoma	Kansas	West Virginia
Kentucky	South Carolina	Louisiana	Wyoming
Mississippi	Tennesee	Montana	
Nevada	Washington		

Source: Energy Information Administration, 1988.

The Energy-consumption Sectors

NCSL provides specific background information from the four traditional energy-consumption sectors (industrial, transportation, residential, and commercial) for each of the six states profiled in this report. Energy consumption for each sector varies significantly from state to state.

Figure 15

Total Energy Consumption by State, 1988

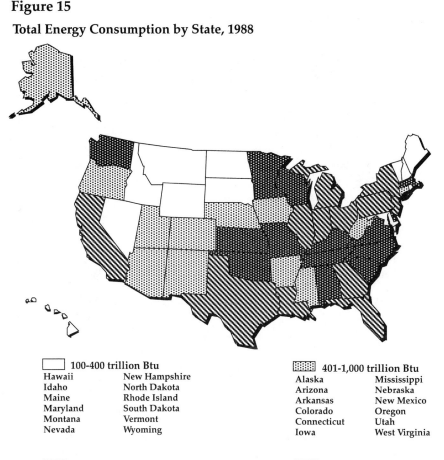

	100-400 trillion Btu			401-1,000 trillion Btu
Hawaii	New Hampshire		Alaska	Mississippi
Idaho	North Dakota		Arizona	Nebraska
Maine	Rhode Island		Arkansas	New Mexico
Maryland	South Dakota		Colorado	Oregon
Montana	Vermont		Connecticut	Utah
Nevada	Wyoming		Iowa	West Virginia

	1,001-2,000 trillion Btu			2,001-9,000 trillion Btu
Alabama	North Carolina		California	Michigan
Delaware	Oklahoma		Florida	New Jersey
Kansas	South Carolina		Georgia	New York
Kentucky	Tennessee		Illinois	Ohio
Massachusetts	Virginia		Indiana	Pennsylvania
Minnesota	Washington		Louisiana	Texas
Missouri	Wisconsin			

Source: Energy Information Administration, 1988.

The industrial sector consumes the most energy in the United States. **Figure 16** shows how electricity is used in this sector. Variable-speed motors, cogeneration, and heat-pump improvements have changed American manufacturing methods. Since a typical industrial motor consumes more than 10 times its capital cost in energy each year, proven energy-saving industrial programs that increase the efficiency of industrial motors are desirable.[3] Similarly, motors are used in other sectors for air transport, refrigeration, fans, and air conditioner compressors.

Figure 16
Industrial Electricity Use in 1990

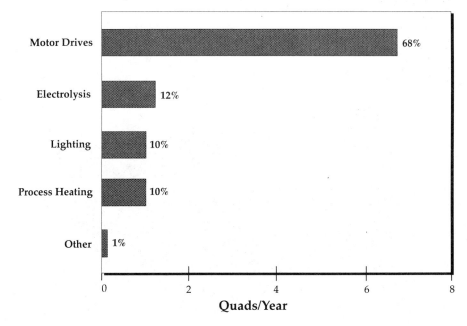

Source: U.S. Department of Energy, Assistant Secretary of Conservation and Renewable Energy, J. Michael Davis, handouts from a speech given to the National Conference of State Legislatures, December 1991, Washington D.C.

There are plentiful examples of industrial energy programs that save millions of dollars in avoided state energy costs. One of the programs featured in this report is New York's Energy Advisory Service to Industry (EASI) program, which hires and trains retired engineers to perform energy audits and make energy-saving suggestions to industrial customers. The New York State Energy Office (NYSEO) says the EASI program saves New Yorkers $64 million each year in avoided energy costs. For the average revenue per kilowatt-hour in the industrial sector by state, see **figure 17**. **Figure 18** shows average revenue per kilowatt-hour for all sectors, by state.

In this study, agriculture is considered part of the industrial sector. The two largest uses of electricity in the U.S. agricultural sector are irrigation pumping (30.8 percent) and dairy farming (18.6 percent).[4] Other uses include materials transport (feed, fertilizer), water heating, crop-drying, and lighting. Farming is an energy-intensive business, using more petroleum than any other U.S. industry, with billions of dollars spent each year on agricultural energy costs. Agriculture accounts for 3 percent of total U.S. energy consumption,

Figure 17

Average Revenue per Kilowatt-hour for the Industrial Sector by State, 1989

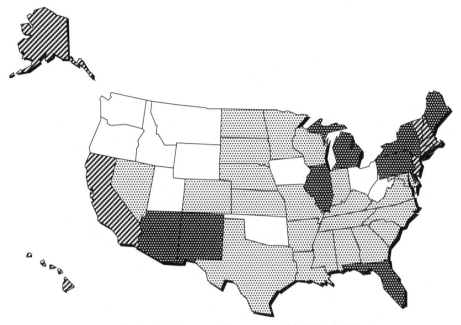

Industrial Average Revenue per kWh is 4.7 Cents

Note: The average revenue per kilowatt-hour of electricity sold is calculated by dividing revenue by sales.

☐ Under 4.0 Cents		▦ 5.1 to 6.0 Cents	▨ Over 6.1 Cents
Idaho-2.6	Utah-4.0	Arizona-5.4	Alaska-7.7
Iowa-4.0	Washington-2.7	Florida-5.1	California-7.1
Montana-3.1	West Virginia-3.6	Illinois-5.4	Connecticut-7.2
Ohio-3.9	Wyoming-3.6	Maine-5.4	Hawaii-6.6
Oklahoma-3.7		Michigan-5.6	Massachusetts-7.3
Oregon-3.3		New Mexico-5.3	New Hampshire-6.7
		New York-5.3	New Jersey-7.2
		Pennsylvania-5.8	Rhode Island-7.5
			Vermont-6.5

▦ 4.1 - 5.0 Cents			
Alabama-4.3	Louisiana-4.3	North Carolina-4.7	
Arkansas-4.9	Maryland-4.7	North Dakota-4.9	
Colorado-4.5	Minnesota-4.1	South Carolina-4.3	
Delaware-4.4	Mississippi-4.8	South Dakota-4.7	
Georgia-4.7	Missouri-4.9	Tennessee-5.0	
Indiana-4.1	Nebraska-4.2	Texas-4.1	
Kansas-4.8	Nevada-4.4	Virginia-4.2	
Kentucky-4.3		Wisconsin-4.1	

Source: Energy Information Administration, Form EIA-861, *Annual Electric Utility Report*, 1989.

with 90 percent of this going to crop production and 10 percent to livestock maintenance. More than 23 million people are employed in agricultural enterprises and 70 percent of the capital assets of all U.S. manufacturing

Figure 18

Average Revenue per Kilowatt-hour for All Sectors by State, 1989

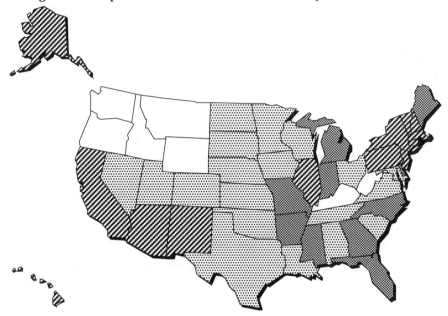

U.S. Average Revenue per kWh is 6.5 Cents

Note: The average revenue per kilowatt-hour of electricity sold is calculated by dividing revenue by sales.

☐ Under 5.0 Cents	▦ 5.1 - 6.0 Cents		
Idaho-3.8	Alabama-5.8	Nebraska-5.5	South Dakota-6.0
Kentucky-4.8	Colorado-5.9	Nevada-5.3	Tennesee-5.4
Montana-4.1	Iowa-5.9	North Dakota-5.7	Texas-5.7
Oregon-4.3	Kansas-5.4	Oklahoma-5.5	Utah-5.8
Washington-3.5	Louisiana-6.0	Ohio-5.7	Virginia-5.9
West Virginia-4.8	Maryland-6.0	South Carolina-5.6	Wisconsin-5.4
Wyoming-4.3	Minnesota-5.3		

▦ 6.1 to 7.0 Cents		▨ Over 7.1 Cents		
Arkansas-6.4	Maine-7.0	Alaska-9.2	Illinois-7.5	New York-8.9
Delaware-6.3	Michigan-6.8	Arizona-7.5	Massachusetts-8.3	Pennsylvania-7.4
Florida-7.0	Mississippi-6.2	California-8.5	New Hampshire-8.4	Rhode Island-8.3
Georgia-6.4	Missouri-6.5	Connecticut-8.8	New Jersey-8.5	Vermont-8.0
Indiana-6.4	North Carolina-6.2	Hawaii-8.1	New Mexico-7.3	

Source: Energy Information Administration, Form EIA-861, *Annual Electric Utility Report*, 1989.

corporations are part of the agricultural sector.[5] Due to these and other factors, agricultural energy-saving programs are valuable components of any state's energy-efficiency strategy. Wisconsin, Iowa, and Nevada agricultural energy programs are profiled in this report.

Energy consumption in the transportation sector is usually the most difficult for states to influence, due primarily to inexpensive gasoline and the independence associated with the automobile. Automobiles and light trucks use the most energy in this sector **(figure 19)**, which is responsible for approximately 75 percent of the oil consumed in a state. While China has one car for every 1,000 people and Russia has one car for every 20 people, the United States has more than one car for every citizen within its borders.[6] Increased fuel-efficiency standards, telecommuting, traffic-light synchronization, and penalties assessed for owning fuel-inefficient automobiles are some of the strategies states use to combat growing energy consumption in this sector. Both Iowa and New York have successful traffic-light programs in place. Washington has a telecommuting demonstration project underway in the Puget Sound area.

Figure 19

Transportation Energy Use in 1989 (Total 22.2 quads/year)

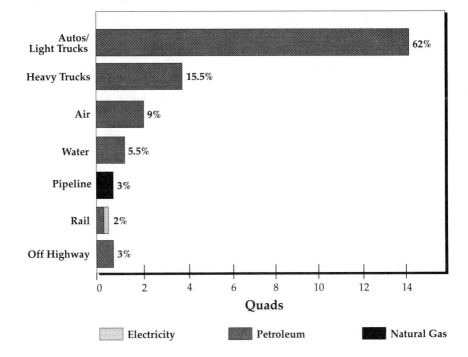

Source: Energy Information Administration, *Monthly Energy Review,* December, 1989.

In April 1992, the Maryland Legislature approved an innovative tax bill, Senate Bill 3, designed to increase fuel effeciency in the transportation sector. The legislation levies a surcharge on consumers who purchase new automobiles with poor fuel efficiency, and provides a credit (rebate) to those consumers who purchase fuel-efficient vehicles. This new "fee-bate" concept is projected to earn $15 million in FY 1995, and much more in later years. For model years 1993 and 1994, the legislation levies a $50 surcharge on consumers who purchase a new automobile with a combined (weighted) highway/city average of less than 21 mpg, and provides a $50 credit to consumers who purchase a new automobile with an average of more than 35 mpg. In 1995, each new automobile will receive a $100 credit for each mile above 35 mpg achieved, while each new automobile that achieves less than 27 mpg will be assessed a surcharge of $100 for each one of those miles. Both the credit and surcharge are limited to one percent of the purchase price of the automobile.

In the residential sector, the largest end use of energy is for heating and cooling buildings. Most of this energy is in the form of electricity; the average revenue per kilowatt-hour in the residential sector by state is shown in **figure 20**. The amount of energy used in a home is directly related to its climatic location. (For energy use in this sector, see **figure 21**.) Proper insulation can reduce heating energy use by 60 percent.[7] Many states have increased efficiency in this sector through traditional weatherization programs funded by the federal government and programs funded by oil-overcharge money. However, some states, such as Washington and Wisconsin, have gone further and through legislation have for years required builders to use energy-efficient building techniques (insulation, double-paned windows, and weatherstripping, for example). Refrigerators and freezers use large amounts of energy as well. Appliance standards like those in Florida are likely to be adopted by other states in the 1990s.

State legislatures have discovered the link between unnecessarily high energy costs and residential housing affordability, especially in low-income housing. Thirty-four states now have voluntary home energy rating systems (HERS) that educate prospective home buyers about energy-efficient building practices. However, some home builders do not use energy-efficient building techniques while building a home, or promote energy-efficiency benefits when selling the home.

The commercial sector uses significant quantities of electricity, with annual electricity sales to commercial customers representing close to 25 percent of total sales.[8] Heating and cooling equipment and windows are especially important in the commercial sector. Accordingly, by switching from traditional incandescent light bulbs to compact fluorescents, commercial energy users in New York are often able to save 40 percent of their annual energy costs.[9] Lighting standards in Florida and New York will be used as models for other states. For the average revenue per kilowatt-hour in the commercial sector by state, see **figure 22.** Commercial lighting—in office buildings, motels, hotels, and groceries—uses significant supplies of energy **(figure 23)**.

Figure 20

Average Revenue per Kilowatt-hour for the Residential Sector by State, 1989

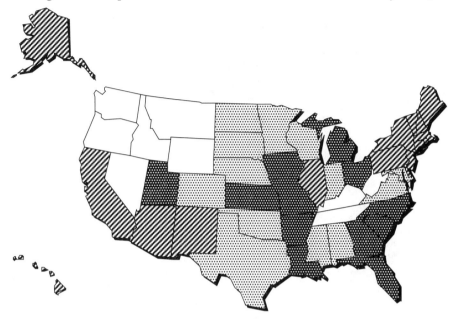

Residential Average Revenue per kWh is 7.6 Cents

Note: The average revenue per kilowatt-hour of electricity sold is calculated by dividing revenue by sales.

☐ Under 6.0 Cents	▦ 6.1 - 7.0 Cents	▦ 7.1 - 8.0 Cents	▨ Over 8.1 Cents
Idaho-4.8	Alabama-6.6	Arkansas-7.8	Arizona-8.8
Kentucky-5.6	Colorado-7.0	Florida-7.7	Alaska-9.8
Montana-5.4	Indiana-6.9	Georgia-7.2	California-9.4
Oregon-4.8	Maryland-6.9	Iowa-7.6	Connecticut-9.6
Nevada-5.7	Minnesota-6.7	Kansas-7.7	Delaware-8.2
Tennessee-5.7	Mississippi-6.8	Louisiana-7.3	Hawaii-9.3
Washington-4.3	Nebraska-6.2	Michigan-7.6	Illinois-10.0
West Virginia-5.9	North Dakota-6.1	Missouri-7.3	Maine-8.5
Wyoming-6.0	Oklahoma-6.8	North Carolina-7.7	Massachusetts-9.1
	South Dakota-6.8	Ohio-7.8	New Hampshire-9.5
	Texas-7.0	South Carolina-7.2	New Jersey-10.1
	Virginia-6.9	Utah-7.4	New Mexico-9.0
	Wisconsin-6.7		New York-10.9
			Pennsylvania-8.9
			Rhode Island-9.0
			Vermont-8.9

Source: Energy Information Administration, Form EIA-861, *Annual Electric Utility Report*, 1989.

Energy-Efficiency Statutes. Section three is a compilation of energy-efficiency statutes for the state. These statutes range from specific legislation to limit the energy consumption of fluorescent light fixtures, to broad legislation such as a mandated statewide energy strategy. The legislation mentioned in this book was chosen by energy experts within each state's

Figure 21
Residential Energy Use in 1990

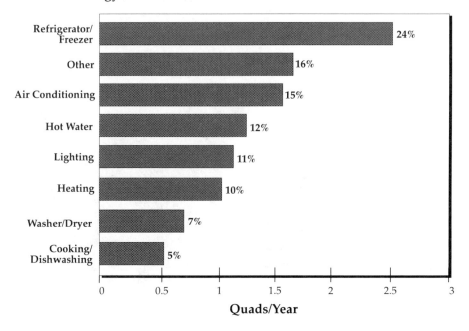

Source: U.S. Department of Energy, Assistant Secretary of Conservation and Renewable Energy, J. Michael Davis, handouts from a speech given to the National Conference of State Legislatures, December 1991, Washington, D.C.

legislative and executive branches. The list is not exhaustive; only significant energy-efficiency legislation was chosen for inclusion. Short summaries of the legislation are included and state legislative service bureaus can be contacted if further details are required (see appendix 10). Please note that the major legislation listed in section three is often included in section four. For example, legislation mandating a state energy strategy in section three would be considered an energy initiative in section four.

Energy-efficiency Programs and Initiatives. Section four describes and evaluates some of the six states' important energy-efficiency programs and legislative initiatives. Noteworthy energy legislation and programs are described at the beginning of this section. Selected state energy programs are evaluated in a matrix toward the end of this section. Footnotes listed within each matrix are explained under a Matrix Notes heading in this section.

Integrated Resource Planning. Also included under this section for each state is an Integrated Resource Planning (IRP) segment. For the purposes of this report, IRP is considered a state energy-efficiency initiative and is

Figure 22
Average Revenue per Kilowatt-hour for the Commercial Sector by State, 1989

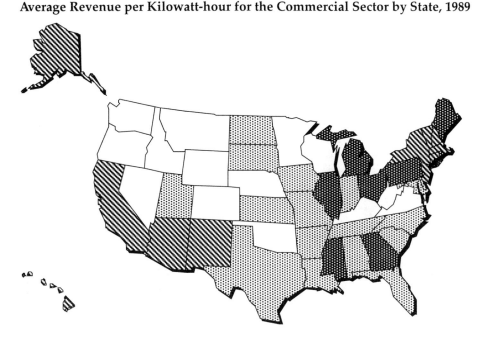

Commercial Average Revenue per kWh is 7.2 Cents
Note: The average revenue per kilowatt-hour of electricity sold is calculated by dividing revenue by sales.

☐ Under 6.0 Cents	▦ 6.1 - 7.0 Cents	▦ 7.1 - 8.0 Cents	▧ Over 8.1 Cents
Colorado-5.8	Alabama-6.8	Georgia-7.3	Alaska-8.8
Idaho-4.3	Arkansas-6.7	Illinois-7.8	Arizona-8.1
Kentucky-5.3	Delaware-6.7	Maine-7.4	California-9.1
Minnesota-6.0	Florida-6.6	Mississippi-7.2	Connecticut-8.7
Montana-4.6	Indiana-6.1	Michigan-7.9	Hawaii-9.2
Nebraska-5.7	Iowa-6.4	Ohio-7.2	Massachusetts-8.1
Nevada-6.0	Kansas-6.5	Pennsylvania-7.8	New Hampshire-8.8
Oklahoma-5.9	Louisiana-7.0	Rhode Island-8.0	New Jersey-8.8
Oregon-4.8	Maryland-6.6		New Mexico-8.2
Virginia-6.0	Missouri-6.5		New York-9.9
Washington-4.1	North Carolina-6.3		Vermont-8.1
West Virginia-5.4	North Dakota-6.5		
Wisconsin-5.9	South Carolina-6.2		
Wyoming-5.2	South Dakota-6.7		
	Tennesee-6.1		
	Texas-7.1		
	Utah-6.8		

Source: Energy Information Administration, Form EIA-861, *Annual Electric Utility Report*, 1989.

addressed in part one of this book. IRP requires utility companies to ". . . share their vision of future energy sources of supply and demand. . ." with state utility planners, while treating as equal demand-side options (energy-efficiency programs) and supply-side options (new power plants). IRP treats saving electricity or increasing the efficiency of existing electricity end users as a supply option for future electricity needs, and often makes new electricity generating plants unnecessary.

Figure 23
Commercial Electricity Use in 1990

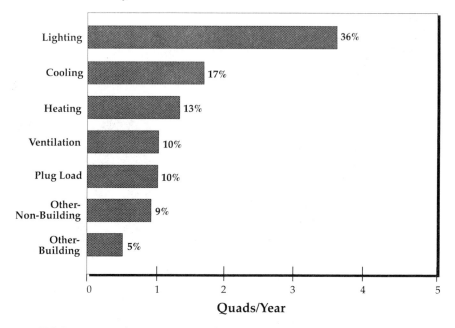

Quads/Year

Source: U.S. Department of Energy, Assistant Secretary of Conservation and Renewable Energy, J. Michael Davis, handouts from a speech given to the National Conference of State Legislatures, December 1991, Washington D.C.

The terminology to describe IRP varies from state to state. Generally, IRP includes the concepts of demand-side management (DSM), conservation and load management (C&LM), least-cost planning (LCP), and energy efficiency. Benefits to be derived from an effective IRP process include increased energy use efficiency; healthier state budgets; and reliable, stable, inexpensive electricity rates for all citizens.

Integrated resource planning for gas utilities is less common than IRP for coal-burning utilities; however, the IRP process is equally applicable to both fuels. At least 16 states require gas utilities to file IRPs. Since natural gas is perceived by many to have substantial environmental and efficiency benefits, utility sector natural gas use is expected to grow substantially in the next decade. Therefore, more state regulatory bodies will likely require natural gas integrated resource plans.

Electricity rates vary substantially between the states, and within the states we have examined for this book. **Table 21** shows the 1989 average revenue per kWh in New York's industrial sector was 5.3 cents, while the average residential revenue in New York was 10.9 cents per kWh. **Figure 24** shows that the average revenue per kWh for all sectors in the United States is 6.5 cents.

Table 21
Average Revenue per Kilowatt-hour by Sector, Census Division, and State, 1989
(Cents)

Census Division State	Residential	Commercial	Industrial	Other[1]	Average
New England	9.2	8.2	6.9	10.6	8.3
Connecticut	9.6	8.7	7.2	12.4	8.8
Maine	8.5	7.4	5.4	10.0	7.0
Massachusetts	9.1	8.1	7.3	10.4	8.3
New Hampshire	9.5	8.8	6.7	12.9	8.4
Rhode Island	9.0	8.0	7.5	7.8	8.3
Vermont	8.9	8.1	6.5	11.3	8.0
Middle Atlantic	9.9	9.0	5.8	8.3	8.3
New Jersey	10.1	8.8	7.2	15.2	8.5
New York	10.9	9.9	5.3	7.7	8.9
Pennsylvania	8.9	7.8	5.8	10.2	7.4
East North Central	8.0	7.2	4.5	6.7	6.3
Illinois	10.0	7.8	5.4	6.8	7.5
Indiana	6.9	6.1	4.1	8.0	6.4
Michigan	7.6	7.9	5.6	8.7	6.8
Ohio	7.8	7.2	3.9	6.0	5.7
Wisconsin	6.7	5.9	4.1	6.3	5.4
West North Central	7.1	6.3	4.4	5.9	5.9
Iowa	7.6	6.4	4.0	6.1	5.9
Kansas	7.7	6.5	4.8	7.5	5.4
Minnesota	6.7	6.0	4.1	6.4	5.3
Missouri	7.3	6.5	4.9	6.9	6.5
Nebraska	6.2	5.7	4.2	6.0	5.5
North Dakota	6.1	6.5	4.9	3.8	5.7
South Dakota	6.8	6.7	4.7	3.8	6.0
South Atlantic	7.3	6.5	4.5	6.0	6.2
Delaware	8.2	6.7	4.4	9.3	6.3
District of Columbia	6.0	6.3	5.1	5.7	5.9
Florida	7.7	6.6	5.1	6.7	7.0
Georgia	7.2	7.3	4.7	7.8	6.4
Maryland	6.9	6.6	4.7	7.9	6.0
North Carolina	7.7	6.3	4.7	6.4	6.2
South Carolina	7.2	6.2	4.3	5.8	5.6
Virginia	6.9	6.0	4.2	5.2	5.9
West Virginia	5.9	5.4	3.6	8.3	4.8
East South Central	6.1	6.3	4.6	5.6	5.4
Alabama	6.6	6.8	4.3	5.6	5.6
Kentucky	5.6	5.3	4.3	4.6	4.8
Mississippi	6.8	7.2	4.8	7.7	6.2
Tennessee	5.7	6.1	5.0	6.8	5.4
West South Central	7.1	6.2	4.1	6.2	5.8
Arkansas	7.8	6.7	4.9	5.7	6.4
Louisiana	7.3	7.0	4.3	6.1	6.0
Oklahoma	6.8	5.9	3.7	5.2	5.5
Texas	7.0	6.1	4.1	5.5	5.7
Mountain	7.2	6.5	4.1	5.2	5.9
Arizona	8.8	8.1	5.4	4.7	7.5
Colorado	7.0	5.8	4.5	7.0	5.9
Idaho	4.8	4.3	2.6	4.6	3.8

Table 21 (continued)
**Average Revenue per Kilowatt-hour by Sector,
Census Division, and State, 1989**
(Cents)

Census Division State	Residential	Commercial	Industrial	Other[1]	Average
Montana	5.4	4.6	3.1	4.5	4.1
Nevada	5.7	6.0	4.4	4.2	5.3
New Mexico	9.0	8.2	5.3	6.1	7.3
Utah	7.4	6.8	4.0	4.6	5.8
Wyoming	6.0	5.2	3.6	6.0	4.3
Pacific Contiguous	7.4	7.8	5.0	4.1	6.7
California	9.4	9.1	7.1	4.5	8.5
Oregon	4.8	4.8	3.3	4.8	4.3
Washington	4.3	4.1	2.7	3.1	3.5
Pacific Noncontiguous	9.5	9.0	6.7	11.9	8.5
Alaska	9.8	8.8	7.7	13.1	9.2
Hawaii	9.3	9.2	6.6	8.5	8.1
Average	7.6	7.2	4.7	6.2	6.5

[1] Includes public street and highway lighting, other sales to public authorities, sales to railroads and railways, and interdepartmental sales. The average revenue per kWh of electricity sold to other consumers may include operation, generation, maintenance, and rental fees for equipment and/or demand and service charges.

Notes: The average revenue per kWh of electricity sold is calculated by dividing revenue by sales. The Form EIA-861 data shown in this publication differ somewhat from Form EIA-826, "Monthly Electric Sales and Revenue Report with State Distributions," estimated data published in the *Electric Power Monthly* and in the *Electric Power Annual* for 1987 and prior years.

Source: Energy Information Administration, Form EIA-861, *Annual Electric Utility Report*, 1989.

The 1989 average revenue per kWh for the states selected in this report are listed below.[10]

State	Average Revenue per kWh
Florida	7.0 cents
Iowa	5.9 cents
Nevada	5.3 cents
New York	8.9 cents
Washington	3.5 cents
Wisconsin	5.4 cents

Bringing energy efficiency to its present status has not been easy. The environmental cost of electricity is one area of conflict. Since the early 1970s, some states have required utilities to evaluate the environmental factors of their future electricity plans; others largely ignore the issue even today. One thing is certain: the methods used to evaluate environmental factors are diverse. Most respondents involved in state utility issues interviewed for this book said that quantifying environmental costs, or externalities, is likely to be the primary issue for utility planners in the next decade.

Figure 24

Average Revenue per Kilowatt-hour for All Sectors by State, 1989

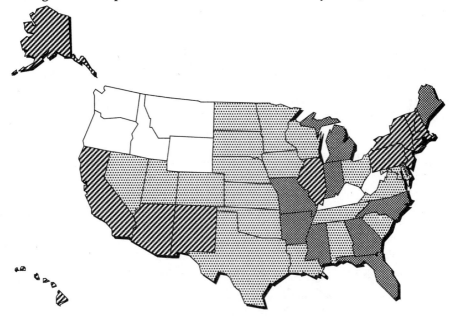

U.S. Average Revenue per kWh is 6.5 Cents

Note: The average revenue per kilowatt-hour of electricity sold is calculated by dividing revenue by sales.

☐ **Under 5.0 Cents**   ▦ **5.1 - 6.0 Cents**

Idaho-3.8	Alabama-5.8	Nebraska-5.5	South Dakota-6.0
Kentucky-4.8	Colorado-5.9	Nevada-5.3	Tennessee-5.4
Montana-4.1	Iowa-5.9	North Dakota-5.7	Texas-5.7
Oregon-4.3	Kansas-5.4	Oklahoma-5.5	Utah-5.8
Washington-3.5	Louisiana-6.0	Ohio-5.7	Virginia-5.9
West Virginia-4.8	Maryland-6.0	South Carolina-5.6	Wisconsin-5.4
Wyoming-4.3	Minnesota-5.3		

▦ **6.1 to 7.0 Cents**   ▨ **Over 7.1 Cents**

Arkansas-6.4	Maine-7.0	Alaska-9.2	Illinois-7.5	New York-8.9
Delaware-6.3	Michigan-6.8	Arizona-7.5	Massachusetts-8.3	Pennsylvania-7.4
Florida-7.0	Mississippi-6.2	California-8.5	New Hampshire-8.4	Rhode Island-8.3
Georgia-6.4	Missouri-6.5	Connecticut-8.8	New Jersey-8.5	Vermont-8.0
Indiana-6.4	North Carolina-6.2	Hawaii-8.1	New Mexico-7.3	

Source: Energy Information Administration, Form EIA-861, *Annual Electric Utility Report,* 1989.

It is not practical to use uniform IRP for all states. Each state legislature is different, and each state has a unique method for evaluating utility plans for the future. Each public utility commission (PUC) has a unique approach to dealing with economic issues, just as each PUC deals with environmental issues differently.

The six states profiled in this book are representative of IRP efforts throughout the United States and illustrate the range of program development from states such as Nevada (with legislated IRP) to states such as Iowa (currently initiating IRP as part of a comprehensive state energy-efficiency strategy).

A number of common factors and issues become apparent when comparing state IRP programs. Although Nevada has legislation to enable its PUC to require IRP programs, the more common impetus has come through intervention in utility rate cases, resulting in regulatory commission orders that instruct utilities to develop IRP programs. A more recent extension of this process is the statewide collaborative conducted by California and by the New England states that combines the efforts of utilities, regulatory commissions, environmental groups, and consumer interest groups in a nonadversarial process to effect a cooperative approach to IRP.

The Matrix. The selection and development of public energy conservation programs is a difficult task for policymakers. Objectives for energy-conservation programs are often multifaceted and fragmented, and the measure of success for these programs is often limited to one or two variables, usually centered on the estimated cost savings or start-up costs associated with the program. NCSL believed a more comprehensive, outside evaluation of state programs would benefit state policymakers. As a preliminary step, a matrix was designed to incorporate both the traditional cost-saving quantitative figures and the qualitative criteria valued by decisionmakers.

The same matrix is used as an evaluation tool at the end of each state section for the programs in that state. Each program listed in the matrix is described in section four of each state section. The energy programs are listed at the left side of the matrix, and criteria used to evaluate the program are listed above and to the right of each program name. The matrix contains footnotes that explain the rating in a given box. Footnotes are brief, one-sentence descriptions that explain the reasoning behind a particular matrix rating. A blank matrix is featured as **table 22**.

The Methodology. Between November 1990 and January 1991, NCSL contacted state energy offices, regional DOE offices, state legislators, legislative staff, PUC staff, state executive branch staff, energy consultants, energy experts employed by federal laboratories, and other energy-related sources. These sources were identified through a formal stakeholder analysis, and were asked to provide a list of states they believed to have the most effective energy-efficiency legislation.

Table 22
Sample Matrix

Energy Program	Cost per kWh (cents/kWh)	Annual Savings ($)	Implementation (easy, moderate, difficult)	Environmental Benefits (low, medium, high)	Start-up Costs (low, medium, high)	Fiscal Fairness (+,−)	Probability of Success (low, medium, high)	Benefit to Cost Ratio (Benefit/Cost)	Limitations or Special Circumstances	Confidence (low, medium, high)

Once the states were selected, sources were contacted within each state who could provide information on the state's energy-efficiency legislation and energy programs and were asked for formal program evaluations, if available; however, with the notable exceptions of New York and Washington, few were provided. Much of the information sent to NCSL was in the form of annual reports (often required by legislation) with no formal benefit/cost analysis. Unfortunately, it appears that most states have neither the staff nor the money to conduct extensive program evaluations. Program evaluations performed by objective, outside sources were virtually nonexistent. When possible, NCSL staff interviewed energy-program managers and related staff and selected programs that would be of interest to legislators and legislative staff and attempted to fit the programs to the matrix. While all of the qualitative judgments, and some of the quantitative calculations about each state's energy programs, are the authors', state energy office personnel and energy program managers from each state were given the opportunity to fill in the matrix as they thought appropriate. With the exception of Florida, all state energy offices contributed significant time toward this task.

The matrix is not intended to replace a formal evaluation for any of the programs described; a formal program evaluation for only one of the energy programs mentioned here could easily be longer than this book. This matrix is designed to be used by state policymakers as a quick reference tool on which to base future decisions about energy programs. NCSL's goal was to identify unique state energy programs funded by oil-overcharge money, the federal government, or the state; document the cost savings, if possible; and add information to the evaluation that could be used by state legislators and energy staff to evaluate the programs further.

Again, the degree of success of a specific energy program is a product of the degree to which a program can meet a full range of public-policy goals. Energy program decisions are not made solely on a one-dimensional cost calculation. New criteria have been added to the one-dimensional cost analysis for a more valuable evaluation. The following criteria and definitions can assist public policymakers to minimize the risks of new energy programs.

Cost Per kWh (cents/kWh). New sources for generating electricity, including nuclear plants and coal powered plants, are expensive to build. It is usually estimated that to construct these new alternative sources of energy, state ratepayers will spend close to $.05 or $.06 ($1990) per kWh for electric power from these new supply sources. The ultimate goal of many state energy programs is to reduce the need for new electrical energy sources. The cost of obtaining new electricity resources through improved energy efficiency and implementation of new energy-saving programs is often less than this figure. For example, to obtain new electricity resources through insulation measures in new, electrically heated homes in Washington may cost as little as

$.02 to $.03 per kWh.[11] This is less than half the cost of new traditional electrical generating resources, and would be cost-effective. Many of the state energy programs featured in this book will not have a cost per kWh figure in the matrix, because the programs did not provide information about cost and/or savings. Where accurate data were available for programs not directly related to electricity production, cost per kWh conversions were made. For example, gallons of gasoline saved by traffic-signal-sequencing programs were converted into cost per kWh figures.

Annual Savings. The annual savings criterion is expressed in 1990 dollars unless specified otherwise, and describes the savings directly attributable to the energy program cited. Many states that report savings of a given program were unsure about annual savings; in such cases, savings were estimated. Such numbers are accompanied by an "E." If savings were actually measured, the number is accompanied by an "M."

Implementation. Three implementation values were assigned—"easy," "moderate," or "difficult." They relate principally to administrative issues of effectiveness in a given state energy program. An "easy" score indicates that the program can be implemented with minimal effort, minimal problems, or a combination of both. Such a score implies minimum expected opposition to a new energy program and minimum change to existing state agencies. A "moderate" score implies normal or average implementation issues can be expected—some change is necessary, but not a significant amount. This score might imply that another state could expect to achieve similar results in the same general timeframe used by the sample state. A "difficult" score implies that possible administrative or procedural reforms may be necessary prior to implementation. Such a score may also imply that there are numerous obstacles to successful implementation. It might also mean that significant time requirements are included as part of the program—it may take years to set up a similar program. Please note that state policymakers interviewed for this report who rated some programs "difficult" often said that the substantial energy savings resulting from a program, once it became operational, justified the difficult implementation of the program.

Environmental Benefits. At a meeting of state legislators in December 1990 in Washington, D.C., the late Bill Hoyt, chairman of the New York State Assembly Energy Committee, commented on the environmental ramifications of energy decisions: "With very few exceptions, energy decisions are environmental decisions." In a 1991 survey of 2,000 American adults across the country, Golin/Harris, a U.S. public relations firm, and the Angus Reid Group, found that 74 percent of Americans believe government should keep environmental protection a priority, even if it means slower economic growth.[12]

For the purpose of this book, environmental benefits are considered primarily a reduction in emissions, from both stationary and mobile sources. Since reducing energy consumption through improved energy efficiency is the goal of many state energy programs—and reduced electricity consumption usually means less air emissions—this criterion is generally related to emissions reductions. As a result, programs responsible for reductions of automobile emissions receive favorable ratings in this category. There are exceptions, of course. For example, due to its small size, Nevada's photovoltaic irrigation demonstration program does not significantly reduce total emissions, but since it keeps cattle away from sensitive riparian habitat, it deserves a "high" environmental benefit score.

There are also three categories for the environmental-benefits criterion—"low," "medium," or "high." While most energy-saving programs will result in significant, long-term, environmental benefits in the form of reduced emissions from electricity production, there are substantial lag-time and monetary environmental-benefit differences. For example, a program that provides energy audits to industry, and in the short term results in decreased emissions from power plants, is given a "high" ranking, since the benefits are immediate and measurable. However, a program that provides workshops on energy-saving building technologies with real, but uncertain, long-term benefits might be given a "low" ranking for providing environmental benefits, since benefits are not measured and would tend to take place over a long period of time. Quantifying environmental benefits is an inexact science, at best; however, the quantification of environmental benefits (and costs) is becoming an accepted and reliable policy tool.

A "medium" ranking might result when a program has both positive and negative environmental consequences, but the positive consequences outweigh the negative consequences. For example, the Wisconsin Conservation Corps (WCC) trains underprivileged inner-city youth in Milwaukee to install energy conservation measures in homes and businesses. The WCC's work is carried out primarily in the suburbs. WCC workers travel to the suburbs daily, which increases vehicle-miles traveled (VMT) and emissions. However, the energy-conservation measures installed by WCC outweigh the increase in VMT, for a "medium" rank. A "low" ranking would imply that environmental benefits are not significant, are questionable, and occur over the long term.

Start-up Costs. The initial costs to start a new energy-saving program are important in today's economic climate. With two-thirds of the states experiencing revenue shortfalls—amounting to several billion dollars in some cases—the initial cost of a program is an important criterion. Start-up costs are rated as "low," "medium," or "high." A "low" ranking implies start-up costs of less than $100,000. A "medium" ranking implies start-up costs of between $100,000 and $500,000, and a "high" ranking implies start-up costs greater than $500,000.

Fiscal Fairness. Fiscal fairness is rated as either a plus (+) or minus (-), and is designed to rate the equity effect of a given energy program. This criterion is similar to the "no-losers test." While open to debate, this criterion is still important to state policymakers. Its intent is that benefits derived from a given program accrue equally to all citizens of the state; however, this is not always the case. For example, a program that provides energy audits to industrial clients and is funded through utility rate surcharges would receive a (-), since all citizens theoretically pay for the program through their monthly utility bills, but only industrial clients benefit directly from the program. Benefits accrue to all citizens through reduced emissions from reduced electricity demand; however, only the specific, direct benefits attributed to the program in question are of interest for this study. A program that provided utility rate rebates to all customers, or that allows all customers to participate in the program, would receive a (+), since the program did not exclude a substantial portion of the public, even though all citizens do not pay utility bills. Most statewide traffic-light-synchronization programs would receive a (+) because the programs are available to all cities and all state citizens benefit from the program through reduced gasoline consumption, decreased automobile emissions, and lower monthly gasoline bills.

Benefit-to-cost Ratio. This criterion is nearly always important to state energy policymakers. To arrive at a simple benefit-to-cost ratio, the benefits of the program are divided by the costs of the project over its lifetime. For example, if an energy-saving program cost the state $100,000 to administer and resulted in $1 million in avoided energy costs (the benefit over the life of the program), the program would have a benefit-to-cost ratio of 10 to 1 (or $1 million divided by $100,000). Energy program benefits vary significantly between programs profiled in this report. For example, benefits derived from a telecommuting project last only as long as the program is in place, whereas benefits derived from new insulation levels required by a new residential building code can be expected to last more than 30 years from the date of installation. A program that pays for itself would have a benefit-to-cost ratio of at least 1:1. In some cases, the benefit-to-cost ratio has been estimated (E), and in other cases the ratio was actually measured (M).

Probability of Success. Ranked as "low," "medium," or "high," the probability-of-success criterion refers to the probability that a program will produce similar results in another state, if that state chose to implement the program using the same general approach. This criterion, admittedly the most subjective, also takes into account how many similar programs have been implemented in other states. For example, a successful residential oil-furnace rebate program in New York might be rated "low," even though the program had a very high benefit-to-cost ratio. The low ranking would be appropriate because there are a disproportionate number of residential oil-burning furnaces in New York. Other states therefore could not expect to achieve the same results, savings, or participation rate in the same program as New York. Southern states—where residential oil-burning furnaces typically are not used—would need to know that New York has a disproportionately high number of oil-burning furnaces before considering the benefits and costs to implement a similar program.

Limitations or Special Circumstances. This criterion allows notation of any peculiarities, problems, or special circumstances associated with a given program's evaluation. In most cases, the evaluation for the program in question was prepared by the program officer, thus allowing a natural bias to enter the evaluation. Questions of program evaluation methodology would also be noted here; for example, a statistically insignificant sample size or survey response size that would not warrant the conclusions made by the program evaluator. Special circumstances surrounding the success or failure of a program, or inconsistencies found in program reports, might also be noted.

Confidence. Ranked as "low," "medium," or "high," this subjective criterion provides an estimate of the "confidence" in the methodology, data, or sources of information provided for each program. This criterion is used by the Western Area Power Administration (WAPA) for similar purposes. Since virtually all the program information reviewed for this report was obtained from respondents vested in portraying their program positively, this is an important criterion. Some detailed program evaluations were prepared by six staff members, while other programs identified for this report have never been evaluated formally. A "high" rating is given when a formal program evaluation or a similar document for the energy program in question was received directly from program officials. A "medium" rating is given when supporting documents, such as annual reports and other related program information, were received from program officials. A "low" rating is given when little written information was received on the program or there was minimal input from program officials familiar with the program.

Future Direction/Legislative Priorities. Section five contains a brief summary of legislation or actions pertaining to energy efficiency that the state expects to address in the near future. Both legislative and executive branch initiatives were solicited from each state for this section. New bills as well as proposed or reintroduced legislation are mentioned here.

Chapter Notes

1. New York State Energy Office, Draft New York State Energy Plan, V (May 1989): 1.

2. New York State Assembly Energy Committee, "New York's Persistent Energy Crisis: The Consequences of Overuse and the Availability of Power Through Energy Efficiency" (January 24, 1991): 8.

3. Western Area Power Administration (WAPA), DSM Pocket Guidebook, 2: Commercial Technologies (April 1991): 79.

4. WAPA, DSM Pocket Guidebook, 3: Agricultural Technologies (April 1991): 1.

5. Raymond P. Poincelot, "Toward a More Sustainable Agriculture" (Westport, Conn.: AVI Publishing Company, Inc., 1986): 1.

6. Special Report, Energy and the Environment, *Battelle Today*, 67 (July 1991): 4.

7. WAPA, DSM Pocket Guidebook, 1: Residential Technologies (April 1991): 7.

8. WAPA, 2: 1.

9. The late Bill Hoyt, New York Assemblyman: personal communication, November 1991.

10. Energy Information Administration, Annual Electric Utility Report, Form EIA-861, Electric Sales and Revenue, 1989.

11. Richard W. Byers, "Cost-Effectiveness of Residential Building Energy Codes: Results of the University of Washington Component Test and the Residential Standards Demonstration Program" (Washington State Energy Office, 1989): 5-1.

12. *Sierra Club National News Report* 23, 14 (July 26, 1991).

FLORIDA

F lorida produces only 1.8 percent of its energy needs within its
borders. Florida residents spent $17.2 billion in 1988 and $18.9
billion in 1989 for energy in its various forms.[1] In planning its
energy future, Florida has special problems. Florida's explosive population
growth in recent years—more than 700 new residents enter the state each
day—strains energy supplies.[2] The state is especially dependent on petroleum
for the transportation needs of its citizens, and the tourists who annually
pump millions of dollars into the state economy. Florida is one of the most
automobile-dependent states in the country. With an infrastructure that
cannot be easily connected or efficiently served by mass transit systems,
Florida's transportation system and tourist-related economy are tied to the
automobile and petroleum, making the state especially vulnerable to
petroleum shortages and price increases. Florida cities are characterized by
energy-intensive, sprawling suburbs connected by strip-developed highways.[3]
Florida relies on petroleum for nearly 50 percent of its total energy
consumption, far above the national average. With petroleum prices cheaper
now than in 1980, this figure is unlikely to decrease in the near future. In 1988,
Florida rated fourth in petroleum consumption behind Texas, California, and
New York.[4]

The citrus industry makes agriculture a major economic activity and,
consistent with substantial population increase, construction is a major Florida
industry. Due in part to the expanding construction industry, commercial
buildings use more energy than ever. To counter this trend, the Florida Solar
Energy Center's Building Design Assistance Center (BDAC) works with the

Governor's Energy Office (GEO) and sophisticated computer programs to review the design of schools, manufacturing facilities, office buildings, and libraries. Money going to this effort is well spent, since more than 100 million new square feet of commercial office space is added each year in Florida.[5] Florida's residential building code is considered one of the most energy-conscious in the country.

The Florida Solar Energy Center (FSEC) was created by the Legislature in 1974 to research alternative energy technologies, to ensure the quality of solar energy equipment in Florida, and to educate residents about energy options. The FSEC has a number of contracts with the GEO and is supported by the Florida electric utility industry, state government, the federal government, and energy/industry research organizations. Coordinated through the FSEC and the Florida Solar Energy Industries Association, specific state efforts will be made in 1992 to encourage greater use of photovoltaics in both the residential and commercial sectors through outreach efforts and demonstration projects.[6]

State policymakers are committed to meet future energy demand with renewable resources. Florida is unique among the states selected for this study because of its development of, and potential to develop, renewable energy. Florida does not have Washington's hydroelectric capacity nor Nevada's arid climate, but it does have a long growing season, high humidity, and warm temperatures. Low-cost fossil fuels have dominated the energy scene for years, and are responsible for the relatively slow advance of renewable energy sources.

In 1992, the Florida Legislature will attempt to identify and eliminate the economic barriers associated with solar energy applications. The Legislature supports the expansion of the solar industry within the state. One minor program supported by the Legislature is the Florida Solar Users Network, which provides a toll-free hotline and assistance to solar system owners and related businesses. The hotline allows callers to locate qualified technical service personnel and solar sales offices and to determine licensing requirements for specific geographic areas.

Biomass is a leading contributor in Florida's renewable energy picture but does not account for a significant portion of the total energy supply. Most of Florida's biomass programs focus on the conversion of municipal solid waste (MSW) to methane gas and compost. Florida's water table is relatively close to the surface, which often prohibits safe and economical landfill disposal of waste. As a result, waste-to-energy projects are attractive resource options for the state. In 1988, 10 Florida facilities recovered energy from municipal solid waste, producing a total of only 19 trillion Btus, or less than one-half of 1 percent of Florida's total energy supply **(figure 25)**.[7]

Figure 25

Florida Energy Use by Source, 1988 (2,928 Total Trillion Btus)

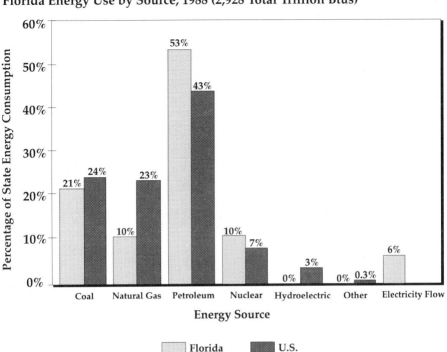

Source: Energy Information Administration, State Energy Data Report, Consumption Estimates, 1960–1988, DOE/EIA-0219, 1988.

Figure 26 shows that Florida's energy consumption increased more than 5 percent between 1987 and 1988. Florida residents spent 8.3 percent of total income on energy in 1988; the national average is 3.3 percent.[8] The GEO explains that the increase occurred due to a 5.7 percent rise in personal income, a 5.1 percent reduction in energy prices, and a 3 percent population increase. Florida's population growth is expected to continue on this trend. While Florida's population increases, its production of both crude oil and natural gas has steadily decreased since 1978 as the amount of recoverable reserves in existing fields dwindled. In 1989, Florida ranked 19th in the nation in crude oil production and 23rd in natural gas production.[9]

In 1980, the Florida Legislature enacted the Florida Energy Efficiency and Conservation Act (FEECA) in recognition of the importance of energy conservation in solving the state's energy problems. The act required the Florida Public Service Commission (FPSC) to set two statewide goals: to reduce the growth rate of both electric energy consumption and weather-sensitive peak electric demand and to reduce the oil consumption level of

Figure 26

Total Annual Energy Consumption, Florida, 1970–1988

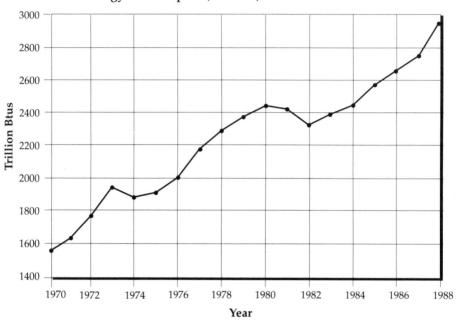

Source: Energy Information Administration, State Energy Data Report, Consumption Estimates, 1960–1988, DOE/EIA-0214, 1988.

electric power plants. With the passage of this act, Florida became one of the first states in the nation to have a legislatively mandated electrical energy-conservation policy.[10] Florida utilities now use less petroleum than in the 1970s. State utilities used petroleum for 17 percent of its energy supply to produce electricity in 1988, compared to the 1972 high of 58 percent.[11] This decrease in petroleum use is due to an increased use of coal to produce electricity. Florida's total energy consumption between 1970 and 1988 is shown in **figure 26**. In 1989, Florida ranked fourth nationally in electricity sales.[12]

Florida is also known for its appliance efficiency standards. Legislation passed in 1987—the Florida Energy Conservation Standards Act (Chapter 87-271, Laws of Florida)—requires the Department of Community Affairs to adopt or modify these standards for new appliances sold in the state after January 1, 1988, for showerheads; after January 1, 1989, for lighting equipment; and after January 1, 1993, for refrigerators and freezers.

In mid-1991, Florida's governor implemented the U.S. Environmental Protection Agency's (EPA) Green Lights Program by cabinet initiative that covered lighting in all state government buildings. With the Green Lights

Program, older, inefficient lighting systems were replaced with more efficient, improved lighting technology. The substantial energy savings that resulted generated interest in other energy-efficiency measures for state government. In October 1991, a broader executive order (no. 91-253) called for more energy-efficiency measures and a 30-percent reduction in state agency energy use by the end of 1994 (see section four).

Energy Consumption by Sector

Generally speaking, energy consumption is increasing in Florida's residential, transportation, and commercial sectors, while consumption in the industrial sector holds steady or decreases. Much of this consumption is explained by the increase in the state's population, economic expansion, and low energy prices.

The Industrial Sector

Florida's industrial sector accounted for only 15 percent of total energy consumption in 1988 (**figure 27**), the least of all states selected for this study. With the exception of agriculture—the citrus industry is a major energy user—Florida is not known for its industry. As part of the industrial sector, agriculture uses a tremendous amount of Florida electricity. Irrigation

Figure 27

A Six-state Comparison of Energy Use by the Industrial Sector, 1988

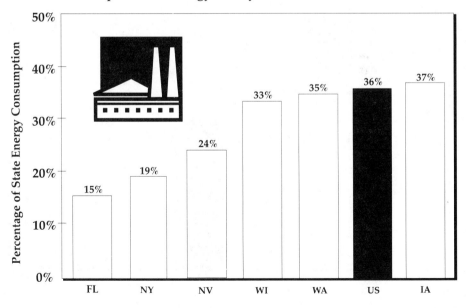

Source: Energy Information Administration, State Energy Data Report, Consumption Estimates, 1960–1988, DOE/EIA-0219, 1988.

and traditional pesticide and fertilizer applications are energy-intensive procedures. The state supports a number of energy-saving agricultural programs, including no-till planting and low-energy grazing. No-till planting saves energy, since only 10 percent of the soil is disturbed and only three or four trips are made across a field, as opposed to the eight or more trips normally required. No-till planting also keeps soil deterioration at a minimum by planting directly into the stubble of a previous crop with a simultaneous application of pesticide. Low-energy grazing encourages Florida livestock producers to graze on native lands, rather than transporting feed or the livestock to a different location, thus saving considerable energy and feed costs. Feed cost is generally the major expense for livestock producers; transportation of both the livestock and feed is a significant expense, as well.

The Transportation Sector

Florida's transportation sector accounts for approximately 40 percent of total energy consumption (**figure 28**). The state has no mass transit programs, and the automobile, inexpensive gasoline, and expanding population contribute to the high consumption rate. During the 1970s, the transportation sector accounted for approximately 55 percent of all petroleum

Figure 28

A Six-state Comparison of Energy Use by the Transportation Sector, 1988

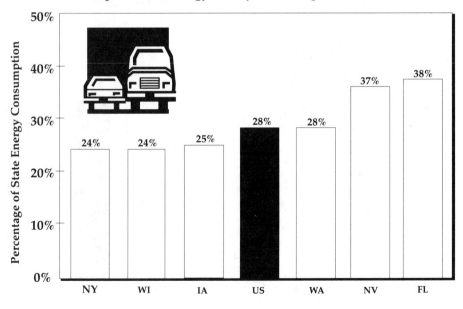

Source: Energy Information Administration, State Energy Data Report, Consumption Estimates, 1960–1988, DOE/EIA-0219, 1988.

consumed in the state; in 1988, this ratio increased to 72 percent. The major cause of this change was the mandated requirement that electric utilities shift from petroleum to coal. The state has committed resources to traffic light synchronization and to a growth management policy to ensure the state's road building programs do not aggravate growth management problems by encouraging even more automobile use.

The GEO, working closely with the Florida Department of Transportation (FDOT), used oil-overcharge funds to support five major areas in the transportation sector—traffic-light synchronization, traffic operations studies, ride-sharing, bicycle and pedestrian projects, and energy-efficiency promotional activities.[13] Ride-sharing projects are marketed as high-visibility public/private partnerships, and in 1989 the GEO funded full-time bicycle coordinators in 14 competitively selected metropolitan areas. A statewide comprehensive pedestrian planning process is ongoing. In an effort to educate fleet operators about the benefits of alternative fuels, the GEO will sponsor demonstration projects in 1992 on methanol, ethanol, hydrogen, compressed natural gas (CNG), and electric or solar powered vehicles.[14] More than 76 million gallons of ethanol-enriched motor gasoline were sold in 1989, but represented only 1.2 percent of total gasoline sales.[15]

Figure 29

A Six-state Comparison of Energy Use by the Residential Sector, 1988

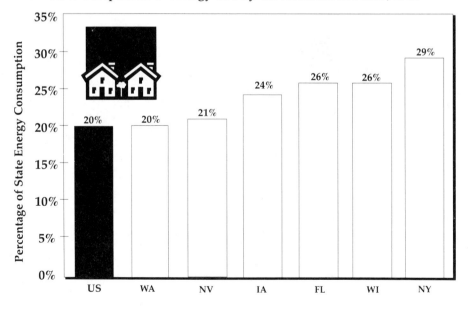

Source: Energy Information Administration, State Energy Data Report, Consumption Estimates, 1960–1988, DOE/EIA-0219, 1988.

138

The Residential Sector

The residential sector consumes 26 percent of total energy in Florida (**figure 29**), although building codes ensure that new residences under construction are energy-efficient. For more than a decade, all jurisdictions in the state have been required to meet the Florida Energy Efficiency Code for Building Construction (ss. 553.87-906 F.S.). Most of Florida's residential energy programs are managed by investor-owned and public utilities. The FPSC requires utilities to offer residential energy audits to their customers. The FPSC addressed energy efficiency in the residential sector until 1989 through the Residential Conservation Service (RCS). Developed by the DOE, the RCS no longer exists.

The Commercial Sector

Florida's commercial sector consumes 22 percent of the state's total energy (**figure 30**). Tourism makes the state's hundreds of hotels and motels primary consumers of electricity. Florida also has a number of energy-intensive commercial fisheries and the associated number of commercial fishing boats, which also consume fossil fuels. The state has started a Small Business Development Center, through a contract with the GEO, to assist

Figure 30

A Six-state Comparison of Energy Use by the Commercial Sector, 1988

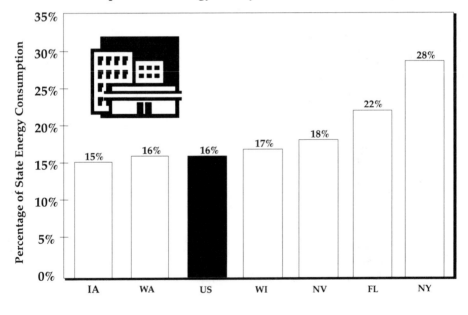

Source: Energy Information Administration, State Energy Data Report, Consumption Estimates, 1960–1988, DOE/EIA-0219, 1988.

Florida's small businesses with ongoing energy management, including auditing, accounting, and innovative financing of energy-conservation measures. Between January 1988 and mid-1989, more than 150,000 new small businesses were started in Florida.[16] Due to the population influx, the commercial sector holds potential for significant energy-efficiency gains in the next decade. The FPSC and the GEO provide technical assistance and training workshops for builders and contractors. The commercial section of the state building code was recently updated. The GEO started a $3 million loan fund for small commercial businesses, and hotels, motels, shopping malls, and restaurants will be encouraged to join state government in the pursuit of energy efficiency in 1992.[17]

Energy-efficiency Statutes

Article VII, Section 3, Paragraph (d) of the Florida Constitution permits an ad valorem tax exemption to a renewable energy source device and to the real property on which such device is installed. The renewable energy device tax exemption is limited to the cost of the device and a 10-year time period.

Article XII, Section 19 of the Florida Constitution provides an effective date of January 1, 1981, for the amendment cited above.

Paragraph 18.10 (3) (c), F.S., 1990 Supplement. Permits the state Board of Administration to invest state funds in loans for a residential energy conservation program in certain circumstances. As long as the Legislature approves, the interest rates are not less than U.S. treasury bill rates, and the funds are not retirement trust funds, the FPSC may devise a plan for the Florida state Board of Administration that allows the state to invest in a residential conservation program the same way the state would invest in any other program. All earnings shall be credited to the general revenue fund.

Section 163.04, F.S., 1989. Prohibits the enactment of any general or special law or adoption of any ordinance that prohibits the installation of any energy device based on renewable resources. The legislative intent of this provision is to encourage the development and use of renewable energy resources to conserve and protect the value of land, buildings, and resources by preventing the adoption of measures that have the ultimate effect, however unintended, of driving the costs to own and operate commercial or residential property beyond the capacity of private owners to maintain. The legislation specifically seeks to protect the use of clotheslines and solar collectors, and does not apply to apartments, condominiums, or cooperatives.

Subsection 187.201 (12), F.S., 1990 Supplement. Establishes the energy goal and policies of the State Comprehensive Plan. This legislation sets one primary energy goal for Florida: To reduce state energy requirements through enhanced conservation and efficiency measures in all end-use sectors, while simultaneously promoting increased use of renewable energy resources (see section four).

Subsection 196.012 (14), F.S., 1989. Defines renewable energy source device for purposes of property tax exemption. This legislation defines renewable energy source devices as any equipment which, when installed in connection with a dwelling unit or other structure, collects, transmits, stores, or uses solar energy, wind energy, or energy derived from geothermal deposits. Among other devices, the legislation lists solar energy collectors; storage tanks; rockbeds; thermostats and other control devices; heat exchange devices; roof ponds; pumps and fans; freestanding thermal containers; pipes, ducts, refrigerant handling systems, and other equipment used to interconnect such systems; windmills; and wind-driven generators.

Section 196.175, F.S., 1989. Fixes the amount of the property tax for property upon which a renewable energy source device is sited. Improved real property with a renewable energy source device installed on it is entitled to a tax exemption not greater than the lesser of the assessed value of the real property less any other exemptions under Florida law; the original cost of the device, including the installation cost, but excluding the cost of replacing previously existing property removed or improved in the course of installation; and 8 percent of the assessed value of such property immediately following such installation. The device must be operative throughout a 12-month period before application can be made for this exemption.

Section 235.212, F.S., 1990 Supplement. Mandates the inclusion of passive design elements and low-energy usage features in the construction of new educational facilities. In addition to requiring special climate control, low-energy use, and natural ventilation where possible, this legislation also requires each new educational facility for which the projected demand for hot water exceeds 1,000 gallons per day to construct, whenever economically and physically feasible, a solar energy system as the primary energy source for the domestic hot water system of the facility. The solar energy system shall be sized so as to provide at least 65 percent of the estimated hot water needs of the facility.

Paragraphs 235.26 (2) (f) and (g), F.S., 1990 Supplement. Require the inclusion of energy performance indices and the performance of life-cycle cost analyses in adopting a uniform statewide building code for educational facilities. The state Board of Education is not to approve construction plans for new facilities unless these plans include an energy performance index and life-cycle cost analysis. The energy performance index shall be a number describing the energy requirements, per square foot of floor space, under defined internal and external ambient conditions over an annual cycle. The life-cycle cost analysis shall be the sum of the reasonably expected fuel costs over the life of the building that are required to maintain illumination, water heating, temperature, humidity, ventilation and all other energy-consuming equipment in a facility; and the reasonable costs of probable maintenance, including labor and materials, and operation of the building.

Section 255.251, 255.253, 255.254, 255.256, 255.258, F.S., 1989. Contain the provisions of the Florida Energy Conservation in Buildings Act of 1974. The original 1974 legislation recognized the energy inefficiency of existing state buildings and called for minimizing energy consumption in new and existing state buildings. The original legislation encouraged shared savings and cooperative energy projects with the private sector. Later legislation required each state agency to appoint "energy management coordinators" to manage energy consumption in each state agency, in an effort to minimize energy waste. Cogenerated power purchases by state agencies are also covered in this legislation.

Sections 255.252 and 255.255, F.S., 1990 Supplement. Contain provisions of the act cited above, which were amended in 1990.

Section 287.083, F.S., 1989. Defines energy conservation terms for the state Division of Purchasing in the context of commodity purchases for the state. This legislation requires the Division of Purchasing to consider life-cycle cost of commodities purchased by the state. Where federal energy efficiency standards exist, the division is required to adopt standards at least as stringent as the federal standards. The division may consider the acquisition cost of the product, the energy consumption and projected cost of energy over the useful life of the product, and the anticipated trade-in, resale, or salvage value of the product.

Section 287.0834, F.S., 1989. Mandates the use of energy saving devices, equipment and additives certified as such by the EPA on all cars bought or leased for more than one year by the state.

Section 366.075, F.S., 1989. Permits the Public Service Commission to approve experimental and transitional rates for any public utility to encourage energy conservation or efficiency. The application of such rates may be for limited geographic areas and for a limited time period. Note: 1989 legislation (chapter 89-292) scheduled this section to be reviewed and repealed in 1999. This legislation allows the FPSC flexibility to experiment with DSM programs and other demand-oriented initiatives.

Sections 366.80-366.85, F.S., 1989. Contain the provisions of the Florida Energy Efficiency and Conservation Act. Section 366.81 declares the legislative intent of the act is to use the most efficient and cost-effective energy-conservation systems, and to reduce and control the growth rates of electric consumption in Florida. The Legislature charged the FPSC to develop and adopt goals consistent with the legislative intent of this act, and directed the FPSC to require each utility to develop plans and implement programs for increased energy efficiency and conservation within its service area. The legislation specifically discouraged the use of petroleum and encouraged the use of cogeneration, load control systems, solar energy, and other renewable energy sources (see section four).

Section 369.105, F.S., 1989. Contains the provisions of the Florida Youth Conservation Corps Act of 1987, which created the corps in the Florida Department of Natural Resources. The legislative intent of this section was to provide public service opportunities to young men and women, while benefitting the state's economy and environment. Section 369.105 (4)(a)(10) provides for work in the energy-conservation area, including the installation of solar hot water devices in public facilities, the performance of energy audits, and energy-conservation improvements in low- and moderate-income housing.

Section 337.703, F.S., 1989. Provides for additional functions of the Executive Office of the Governor with respect to development of an energy emergency contingency plan and coordination and implementation of federal energy programs delegated to the state. This extensive legislation describes the duties and responsibilities of the Executive Office of the Governor with respect to the development of a state energy policy. Included in these duties are the consideration of alternative scenarios of statewide energy supply and demand for five, 10, and 15 years; preparation of an annual report to the Legislature, 60 days prior to each regular legislative session; development of energy training and conservation programs; promotion of waste-to-energy technologies; promotion of the commercialization of solar energy technology; and the promotion of renewable energy sources in general. (This list is far from exhaustive; see actual legislation for details.)

Section 377.706, F.S., 1980. Establishes the Florida Energy Research and Development Task Force, charged with coordinating, promoting and monitoring energy research within the state, in addition to establishing a priority agenda for energy research consistent with state energy policy.

Section 377.709, F.S., 1989. Directs the Public Service Commission to establish a funding program to encourage the development by local governments of solid waste facilities that use solid waste as a primary source of fuel for the production of electricity. Section 377.709 declares that the combustion of refuse by solid waste facilities to supplement electricity supply not only represents an effective conservation effort, but also represents the environmentally preferred alternative to conventional solid waste disposal in Florida.

Section 420.5085, F.S., 1989. Empowers the Florida Housing Finance Agency, located within the Department of Community Affairs, to provide energy conservation loans to homeowners under certain conditions, which include that the sum of the loans cannot exceed $15,000 for any one homeowner and that the home must be the principal residence of the borrower. The loans finance alterations, repairs, and improvements to the energy efficiency of the property; no loans are allowed for swimming pools, saunas, or recreational facilities.

Section 553.14, F.S., 1989. Contains the provisions of the Water Conservation Act. This act requires that no new building constructed after 1983 shall use a tank-type water closet with a tank capacity in excess of 3.5 gallons, nor use a showerhead or faucet that allows a flow of more than an average of three gallons of water per minute at 60 pounds of pressure per square inch. These requirements also apply to additions or renovations if the cost of the addition or renovation exceeds 25 percent of the value of the existing building. Any person who violates this provision is subject to a fine not to exceed $250.

Paragraph 553.73 (1) (a), F.S., 1989. Contains state minimum building codes that require local governments and state agencies with building construction responsibilities to adopt a building code to cover all types of construction. This legislation created a state minimum building code, which consisted of the following nationally-recognized model codes:

1) Standard Building Codes, 1988 edition, pertaining to building, plumbing, mechanical and gas, and excluding fire prevention;

2) EPCOT Code, 1982 edition;

3) One- and Two-Family Dwelling Code, 1986 edition; and

4) The South Florida Building Code, 1988 edition.

Each local government and state agency with building construction responsibilities shall adopt one of the state minimum building codes as its building code. If the One- and Two-Family Dwelling Code is adopted for residential construction, then one of the other recognized model codes must be adopted for the regulation of other residential and non-residential structures.

Section 553.900-553.912, F.S., 1989. Contains provisions of the Florida Thermal Efficiency Code. Requires thermal efficiency provisions be included in state minimum building codes to provide a statewide uniform standard for energy efficiency in the thermal design and operation of all buildings. A number of building types are exempted, and the legislation required all buildings constructed after March 1979, for which building permits were required, to take into account exterior envelope physical characteristics, including thermal mass; heating, ventilation, and air conditioning (HVAC); service water heating; energy distribution; lighting; energy managing; and auxiliary systems design and selection. Similar requirements under this section exist for renovated buildings. The thermal efficiency code also requires that an energy performance disclosure for all new residential buildings be prominently displayed until the time of sale. A statewide uniform display card is used to indicate the energy performance of the dwelling unit.

Section 553.951-553.975, F.S., 1989. Contains provisions of the Florida Energy Conservation Standards Act. The purpose of this act is to provide statewide minimum standards for energy efficiency in certain products, consistent with state energy-efficiency goals. The standards are designed to reduce Florida's energy-consumption growth rate and the growth rate of energy demand. The Legislature designed these product standards so that Florida consumers would benefit over the long run. Products covered under this act include refrigerators, refrigerator-freezers, freezers, lighting equipment and showerheads. The provisions of this act are explained in section four.

Energy-efficiency Programs and Initiatives

Renewable Resources and the State Comprehensive Plan

In Section 187.201 (12) of the 1990 supplement to Florida's 1989 Statutes, the Legislature adopted as part of its state Comprehensive Plan the energy goals and policies described below.

Goal: To reduce its energy requirements through enhanced conservation and efficiency measures in all end-use sectors, while at the same time promoting an increased use of renewable energy resources.

Policies:

1) Continue to reduce per capita energy consumption;

2) Encourage and provide incentives for consumer and producer energy conservation and establish acceptable energy performance standards for buildings and energy-consuming items;

3) Improve the efficiency of traffic flow on existing roads;

4) Ensure energy efficiency in transportation design and planning, and increase the availability of more efficient modes of transportation;

5) Reduce the need for new power plants by encouraging end-use efficiency, reducing peak demand, and using cost-effective alternatives;

6) Increase the efficient use of energy in design and operation of buildings, public utility systems, and other infrastructure and related equipment;

7) Promote the development and application of solar energy technologies and passive solar design techniques;

8) Provide information on energy conservation through active media campaigns;

9) Promote the use and development of renewable energy resources;

10) Develop and maintain energy preparedness plans that are both practical and effective under circumstances of disrupted energy supplies or unexpected price surges.

Florida's Coordinated Energy Plan

Florida's governor, Lawton Chiles, seems interested in pursuing a new energy-efficiency strategy; in 1991 he began a number of energy initiatives that are supported by the Legislature. In the summer of 1991, a cabinet-level resolution was passed to support the implementation of the EPA's successful Green Lights Program, which provides for replacement of existing state agency lighting with more efficient, energy-saving lighting technologies.

On October 8, 1991, Governor Chiles signed Executive Order 91-253, to make the Florida economy more energy efficient, while conserving resources and maintaining the environment. The order calls for reducing state agency energy consumption by 30 percent—15 percent in 1992, and another 15 percent over the next two years.

Section one of the order requires state agency heads to review the laws requiring energy efficiency in government, including section 255.257, F.S., and to take these steps:

1) Within 60 days, submit an initial report on low-cost and no-cost energy-saving efforts to include behavioral, procedural, and equipment modifications to be initiated during the existing fiscal year (1991-1992).

2) By January 1, 1992, submit for inclusion as a budget amendment for FY 1992-1993 further plans for complete energy audits and three-year retrofit plans for all buildings and facilities to produce the desired 30 percent reduction over the next three years.

3) By January 1, 1992, submit as a proposed budget amendment for FY 1992-1993 a plan to begin use of alternative-fueled fleet vehicles in air quality nonattainment areas.

4) Identify cost-effective opportunities to use commercially available, newly emerging technologies in high-visibility demonstrations.

The governor will propose necessary changes to:

1) Remove any legal or institutional obstacles to implementation of energy-efficiency programs.

2) Provide incentives to state agencies to work for maximum energy efficiency, such incentives to include:

a) Retention of energy savings in agency budgets as discretionary income,

b) Establishment of award programs in all agencies for energy-efficiency suggestions by employees,

c) Productivity wage increases from agency energy savings.

3) Expand options for financing energy-efficiency projects.

4) Set overall energy reduction targets for all agencies, emphasizing accelerated implementation with clear responsibilities, commitment, budgeting, auditing, and reporting.

5) Begin use of alternative-fueled fleet vehicles in FY 1992-1993, with the goal of operating all possible fleet vehicles on alternative fuels.

6) Through the Office of Planning and Budgeting, review all agency plans and budgets to ensure maximum effective implementation of Florida's energy-efficiency goals and report the conclusions to the Legislature and the governor.

The order created the Governor's Inter-Agency Council on Energy Efficiency; each agency head is to designate one member to the council, which will coordinate efforts to increase energy efficiency in state government. The GEO will assist agencies with energy-efficiency planning, and the Florida Department of General Services is requested to:

1) Bring the state's energy-conservation manual for all buildings up to the best and highest current standards for energy efficiency, and ensure that all new state buildings constructed in 1992 and thereafter conform to these new standards.

2) Evaluate all state property-leasing practices for energy efficiency.

3) Develop a procedure for the cost-effective bulk purchase of natural gas and other energy-efficient fuels by state agencies.

4) Supervise and enforce the energy-efficiency operations standards (temperature settings, lighting, systems maintenance) of all state facilities.

In 1992, the governor is expected to establish an infrastructure to support private sector entry into the energy-efficiency area. Energy engineers, architects, financial leaders, and the media will be challenged to lead the private sector by example.

The Florida Energy Conservation Standards Act

On March 17, 1987, President Reagan signed the National Appliance Energy Conservation Act (NAECA) to require manufacturers, as of January 1, 1990, to meet federal energy-efficiency standards for refrigerator-freezers, water heaters, HVAC systems, ranges, washers, dryers, and dishwashers.

Rather than wait for the 1990 federal deadline, Florida became the first state to adopt more stringent appliance standards with the passage of the Florida Energy Conservation Standards Act (FECSA) in July 1987. This law sets the following requirements:

> 1) Showerheads sold after January 1, 1988, must be low-flow models;
>
> 2) Most fluorescent lights sold after January 1, 1989, must use high efficiency ballasts; and
>
> 3) All refrigerators and freezers sold after January 1, 1993, must be 25 percent more efficient than those meeting the standards of the National Energy Conservation Standards Act of 1987.

In its first year (1988), the Florida showerhead standard saved electric consumers an estimated $4.76 million.[18] FECSA is expected to produce savings of $100 million per year by 1996 as the effects of the energy-efficient showerheads, lights, and refrigerators and freezers begin to accumulate. By 2012, these savings are projected to grow to $200 million per year. The cumulative savings in nominal dollars reaches almost $4 billion by 2016 as the last "old" refrigerators and freezers are replaced with efficient models.[19]

In 1988, Florida's fluorescent light ballast standard became a national standard applicable in all states when it was included by the federal government in a modification of the NAECA.

Florida's appliance-efficiency standards are expected to increase energy security as well as yield environmental benefits. Without these standards, the state would need to build another 600-megawatt coal-fired electric power plant in the future. With more efficient appliances, utility power plants will burn less coal each year, saving over one million tons annually. The environmental benefits of this avoided power generation include 14,000 fewer tons of sulfur dioxide emissions per year and 100,000 fewer tons of coal ash and scrubber sludge, respectively, to be buried in landfills.[20]

The Florida appliance-efficiency standards are proving cost-effective for electric consumers and offer a high degree of certainty in load reduction for electric utilities. The electric customers, electric utilities, and Florida's environment benefit from this appliance efficiency legislation. The provisions of this act do *not* apply to:

> 1) New products manufactured in Florida and sold outside the state;
>
> 2) New products manufactured outside Florida and sold at wholesale in Florida for final retail sale and installation outside Florida; and

3) Products designed expressly for installation and use in recreational vehicles or other equipment designed for regular mobile use.

The Local Energy Engineer Program

The GEO recognized that many local governments were reluctant to invest in on-site energy expertise because of overall budget demands. The Local Energy Engineer Program (LEEP) was designed in 1990 to guarantee that an investment in an energy engineer will be offset by actual savings in local government operating expenses. This is not a shared savings program, since the local government retains 100 percent of actual dollar savings.

This program provides a two-year salary guarantee to participating local governments that hire an energy engineer. Participating cities hire an engineer on staff or as a consultant to implement energy-conservation measures for the city government. After two years, if the savings from implementing the recommended energy-conservation measures (suggested by the engineer) do not represent the cost of hiring the engineer, the GEO will reimburse the city. For example, if the projected annual savings recommended by the energy engineer are only $20,000 and his or her salary is $30,000 per year, the GEO will guarantee (pay) the difference of $10,000.

The LEEP is exclusively awarded to units of local government—a county or municipality—that can apply individually or jointly with other local governments. The energy engineer is supervised by the chief elected official(s) and chief administrator(s). LEEP contracts were awarded in 1990 to Fort Lauderdale; Bradenton; Lakeland; Gainesville; and the tri-city area of Cocoa, Rockledge, and Melbourne. Each contract guaranteed a maximum of $60,000 over a two-year period to offset hiring an energy engineer if the city did not realize a like dollar figure in energy savings. As of March 1991, no city had requested payment.

Oil-overcharge funds finance the program, and the GEO contracts with the local governments. Energy and financial savings are documented with verified simple payback calculations established by the GEO. The GEO publishes a formal request for proposal (RFP) to provide guidelines for submitting proposals for LEEP.

The energy engineer is responsible to identify and implement an energy-accounting system and plan of action, identify and implement appropriate energy-conservation measures in local government operations, and incorporate energy-efficient land-use policy in local government comprehensive plans. The primary activities of the energy engineer:

1) Establish a data base on energy consumption in local government buildings, fleet vehicles, street lighting, and sewer and water facilities. The data base must include gas consumption per vehicle by fuel type and cost of fuel.

2) Identify opportunities to reduce energy consumption in local government operations.

3) Develop a program to reduce energy consumption in local government operations.

4) Disseminate energy information to community businesses and residents.

5) Incorporate integrated, energy-efficient land-use criteria in the local government comprehensive plan.

Solid Waste Recycling

In 1988, the Florida Department of Environmental Regulation (FDER) received approval from the Legislature to use $18.5 million in stripper well funds to supplement implementation of the Solid Waste Management Act (SWMA), passed during the 1988 legislative session. The 1990 Florida Legislature appropriated an additional $6 million in oil-overcharge funds to continue this program. The State of Florida is committed to spending $150 million over five years on this program.

This major recycling program is included in this report to underscore the undeniable link between energy use and waste minimization. Energy is needed to produce, transport, and recycle Florida consumer products. Florida's increase in population, its limited landfill sites, and a relatively high water table make waste minimization and waste-to-energy high priorities for the state in the early 1990s. Waste transportation costs can be high and uneconomical for recycling purposes if the recycling center is in a distant location. This program attempts to assist local governments to set up recycling programs.

Funds awarded to local governments through this program are used to purchase recycling equipment, provide solid waste education activities, and to set up local used-oil recycling centers. Paper, glass, plastic, construction and demolition debris, and metals are recycled through local programs. Solid waste education grants available through this program promote recycling, volume reduction, proper disposal of solid wastes, and development of markets for recycled materials. The used-oil grants promote the establishment of at least one public used-oil collection center in each of Florida's 67 counties. With assistance provided by this program, all counties have an opportunity to establish solid waste recycling programs in the next two years.

Florida Solar Users Network

This program, directed at the general public, solar system owners, the solar industry construction trade, and others, was funded at $150,000 in 1990–1991.[21] In Florida, the vast majority of solar energy systems provide domestic hot water. Since 1985, it has been difficult for Florida homeowners to obtain maintenance and repairs for their solar systems. Due to rapidly changing technology, some companies have gone out of business, while others have diversified or abandoned older solar energy products.

As part of the program, a toll-free hotline has been provided by which an owner or prospective buyer of a solar system can obtain free information about different types of solar systems and find a qualified service or sales center in the area. Licensing requirements for each area are also provided through the toll-free number. A consumer newsletter is published, as well as a checklist to help system owners with preventive maintenance and minor repairs. A service training program for the solar industry and related trades has also been established through this program.

Mass Transit Demonstration Program

Florida is experimenting with alternatively fueled transit buses. The Metro-Dade Transit Authority (MDTA) is conducting a demonstration of alternatively fueled buses under the Urban Mass Transportation Authority's (UMTA) Alternative Fuel Initiative, with the objective of establishing optimum operating conditions.[22] The MDTA has purchased a total of 25 buses and divided them into five groups—diesel, particulate trap, methanol, compressed natural gas, and reformulated gasoline—and is maintaining fuel consumption, maintenance, and emission comparisons among operating modes. The target area, Dade County and surrounding counties, is a major tourist and convention center with a population of over 1.8 million.

The MDTA estimates that three million trips per year will be taken by bus instead of car when the additional buses and routes are offered. The alternatively fueled buses will improve air quality in this nonattainment area, as well. The Florida Legislature appropriated $600,000 to supplement the MDTA Alternative Fuel Project. UMTA contributed $3.6 million, and the Metro-Dade local government contributed another $600,000 for this initiative.

The Florida Solar Energy Center

Created by the Florida Legislature in 1974, the FSEC is a statewide research institute. With an annual budget of approximately $7 million, the FSEC conducts research on alternative energy technologies. Research centers on photovoltaic issues, passive cooling, HVAC improvements, power electronics, and hydrogen applications. The FSEC, staffed by engineers, physicists, architects and educators, is located at Cape Canaveral.

The center is supported by approximately $3.5 million from the state. The remainder of the center's budget comes from private and federal contributions. The state's contribution reportedly saves $100 million in energy costs each year for Florida citizens.[23] The University of Central Florida provides administrative support to the FSEC. Most of the FSEC's federal contracts require matching state funds, so legislative support is important.

FSEC accomplishments and major activities include:

1) A photovoltaics research and development (R&D) program focused on system development and integration of utility, residential, and stand-alone applications;

2) R&D on advances in building energy use in hot, humid climates;

3) Research on thermal storage systems;

4) Education and training of energy specialists;

5) Commercial energy-code improvements;

6) Hydrogen production from renewable resources;

7) Solar applications in Florida schools;

8) Environmental and energy education in Florida middle schools;

9) Solid waste education; and

10) Monitoring of commercial heat pumps and conservation retrofits.

Recognizing the importance of public awareness, the FSEC's public information office produces a quarterly newsletter and distributes approximately 120,000 publications annually. Technology transfer is a high priority for the center.

Florida Energy Efficiency Code for Building Construction

Since October 1, 1980, all jurisdictions in the state have been mandated under the Florida Building Codes Act to meet the requirements of the Florida Energy Efficiency Code for Building Construction (ss.553.87-906, F.S.). This code provides thermal and lighting efficiency standards for new and renovated buildings, taking into account the structural envelope, building orientation, hot water heating equipment, and heating and cooling equipment selection. The code is designed to encourage the most cost-effective energy configuration for all structural measures and for heating, cooling, and water heating. All minimum performance and efficiency standards prescribed by the code are established based on life-cycle cost parity. Additional efficiency levels may be included at the discretion of the builder. Major revisions were made to the Energy Code in 1984, in which the stringency of the residential compliance levels was increased by 25 percent. The code is strengthened further every two years, based on extensive research. A commercial building code has also been developed, and is being further refined.

One unique aspect of Florida's Energy Code is its use of an energy performance index (EPI) as a method to rate energy efficiency. The EPI is calculated by determining the ratio of projected annual energy consumption of a residence, compared to a standard for that structure size, and multiplying the ratio by 100. Thus, homes which obtain an EPI of 100 or less comply with the code. The chief benefits of this system are that it provides a method for designing energy-efficient homes, it provides consumers with a rating to compare energy efficiency, and it provides a flexible energy-consumption standard for structures of various sizes in different areas of the state. The University of Florida developed a computer software program that assists building officials and contractors to determine the EPI of residential structures.

The Florida Department of Community Affairs (FDCA), Bureau of Codes and Standards, is responsible for the administration of the energy code. Code compliance certification is handled by local building departments. Reporting forms must be completed by the owner or the contractor for each building permit issued and a copy of the reporting form is mailed directly to FDCA. Instructional workshops conducted by FDCA around the state provide information on code standards and reporting requirements to building inspectors and contractors.

Integrated Resource Planning

Florida performs IRP on an as-needed basis. Rule 25-17-0833 requires that the FPSC periodically review utilities' optimal generation and transmission plans from both statewide and individual utility perspectives. Florida held its most recent planning hearings for investor-owned utilities in May 1991.[24] The FPSC also reviews a utility's plan when determining the need for power plant construction. There are numerous rules and statutes in place that combine to form a semi-formal IRP process.

Numerous statutes (Florida Statutes 366.80-82, 366.051, 403.519, 403.5235-36, 403.501-517) and rules (Florida Administrative Code 2517.051-065, 2517.080-091, 2517.001-015, 2522.080-018, and 2522.075-076) cover areas that range from determining the need for electrical power and biennial planning, to transmission line adequacy and cogeneration pricing. Need determination is separate from the planning hearings.

Public hearings on the IRP process were underway in late 1991. The hearings were organized by the FPSC and the Utilities Subcommittee of the House Committee on Regulated Industries and Licensing.

FPSC staff say that electrical generation is very competitive, with qualifying facilities (QFs) competing against each other and the utilities for the right to supply generation. Cogeneration is projected to provide up to 25 percent of new generation needs by 2000.[25] There are plans to provide approximately 1,600 megawatts (MW) of cogenerated power by 1995, and another 600 MW between 1995 and 2000.[26]

Florida's advance planning hearings take place approximately every two years. The FPSC schedules hearings to examine incentives in the rate-making process for utilities to embark on additional demand-side strategies.

In 1980, the FPSC adopted conservation goals pursuant to the FEECA, which was passed that year by the Legislature (366.82, Florida Statutes). In response to conservation goals, utilities have implemented cost-effective conservation plans and programs and have undertaken actions to reduce oil as a generation fuel. These actions have slowed the growth rate in peak electrical demand in both winter and summer, have slowed the growth in kWh sales of electrical energy and have reduced oil use for electrical production from 77 million barrels in 1981 to 33 million barrels in 1987.

Also in 1980, the FPSC established conservation goals for natural gas utilities.[27] These goals were designed to promote the use of natural gas as a substitute for oil or oil-derived energy when it is cost-effective to do so. Natural gas is used concurrently with electric utility peak periods to provide more efficient use of Florida electricity generation.

The legislative sunset review of the FEECA statute occurred during the 1989 session. The major change resulting from that review was the addition of cogeneration as a conservation measure that will have an important role in meeting the future electricity needs of Florida. The FPSC revised its conservation goals in 1990. In addition, the FPSC ordered (Order Number 22176, issued 11-14-89) all electric and natural gas utilities to submit conservation plans and programs in order to implement the legislative intent embodied in FEECA.

Florida's electric utilities must submit 10-year plans, pursuant to Chapter 186.801, Florida Statutes. Recent 10-year plans show average annual load growth rates of approximately 3 percent, compared to a historical annual average increase of about 4 percent,[28] although Florida's population has increased approximately 3 percent per year since the late 1980s.

The FPSC's conclusions regarding future electric generation and transmission additions are based on the generation plans and determination of need proceedings pursuant to sections 403.501 through 403.517 and 403.519, Florida Statutes, and biennial planning hearings.

Utilities must submit specific end-use oriented programs for commission approval. If a program is determined to be cost-effective, contributes to the conservation goals, and is projected to achieve measurable results, utilities may bill prudently incurred expenses to the ratepayers. These revised plans and programs were recently put in place. Recovery is permitted through a six-month conservation cost recovery "true-up factor," similar to a fuel adjustment, that is applied to all jurisdictional kWh sales.[29]

The major investor-owned electric utilities in Florida are Florida Power and Light Company (FPL), Florida Power Corporation (FPC), Gulf Power Company (GPC) and Tampa Electric Company (TEC). FPL practices IRP in Florida, and uses software developed by the Electric Power Resource Institute (EPRI) to evaluate demand-side options. FPC uses advanced load management programs as part of its demand-side strategy.

A Florida loan guarantee program allowed the FPSC to use up to $5 million in regulatory trust fund money to guarantee loans made by utilities or financial institutions to customers who purchase and install approved, cost-effective energy-conservation measures. This program was discontinued June 30, 1991.

Pursuant to Chapter 403, Florida Statutes, the commission is charged with the responsibility to determine the need for power plants proposed for construction by Florida electric utilities. As an improvement to this process, the commission initiated hearings on load forecasts, generation expansion planning studies, and cogeneration prices. In doing so, the commission sought to implement the legislative mandate of Chapter 366.04(3), Florida Statutes, to exercise jurisdiction over the "planning, development, and maintenance of a coordinated electric power grid throughout Florida to assure an adequate and reliable source of energy for operational and emergency purposes in Florida and the avoidance of further uneconomic duplication of generation, transmission, and distribution facilities."

In 1985, the FPSC changed its generation planning and annual planning from a workshop format to a hearing format, to better address the complicated conservation and cogeneration issues that affected statewide energy consumption.

In 1989, coal accounted for 39 percent of total fuel use in the electric sector, petroleum for 17 percent, nuclear for 14 percent, and natural gas for 12 percent. Purchases of electricity generated in adjoining states, increased by 46 percent from 1988 to 1989, made up the balance. Interstate purchase of electricity increased.[30]

While Florida utilities plan to use more coal for electricity production in the future, there has been opposition to this trend at the local level. After learning that the city of Tallahassee planned to buy more coal-powered electricity from a major utility, local citizens gathered enough signatures to bring the issue to the polls. In 1992, Tallahassee voters passed a referendum that could prohibit future purchases by the city of electricity produced from burning coal.[31]

Table 23
Florida Matrix

FLORIDA Energy Program	Cost per kWh (cents/kWh)	Annual Savings ($)	Implemen- tation	Environ- mental Benefits	Start-Up Costs	Fiscal Fairness (+,-)	Probability of Success	Benefit to Cost Ratio (Benefit/ Cost)	Limitations or Special Circum - stances	Confidence
Energy Conservation Standards Act	Not available	$4.7 million E (1)	Moderate (2)	High (3)	Medium (4)	- (5)	High (6)	N/A (7)	Very cost- effective (8)	Medium (9)
Local Energy Engineer Project	Not available	$120,000 E (10)	Easy (11)	Medium (12)	Medium (13)	+ (14)	High (15)	N/A	High front- end costs (16)	Low (17)

Matrix Notes

These notes correspond to the numbers in the matrix (**table 23**).

1) This figure is in 1988 dollars. Annual savings are estimated to be $100 million per year by 1996, and total savings during 1988–2016 are expected to be $4 billion (see chapter note 10).

2) The implementation of energy-efficiency standards for consumer products, like the implementation of building code standards, can be challenging. However, the administrative burden on state government is relatively moderate for a similar act.

3) Environmental benefits are high. Energy savings and associated decreases in electricity production have already accumulated through the showerhead standards.

4) Start-up costs for this program are less than $500,000.

5) A substantial portion of Florida citizens— notably low-income households and owners of homes built in the early to mid-1980s—are ineligible for direct program benefits. These groups are at a special disadvantage, since low-income households cannot afford to purchase new appliances, and the owners of homes built in the early to mid-1980s will not to replace existing appliances for a number of years.

6) Savings are high and well documented. The act is easy to duplicate.

7) Reliable estimated benefit numbers were available; however, no reliable cost information was supplied to NCSL.

8) Typically, 3 percent of total state energy demand is used to heat water. Therefore, substantial energy savings are possible for any state that promotes low-flow showerheads. Appliance energy-efficiency standards have improved in recent years and, in the long run, appliance standards can result in significant savings.

9) No program evaluations were available for this analysis. However, an economic analysis of the act was available (see chapter note 18).

10) NCSL was unable to locate any annual savings figures. This conservative figure was calculated by multiplying the number of energy engineers hired annually (4) by the engineers' annual salaries ($30,000).

11) Administrative requirements for this program are minimal. The RFP procedure is easy to design and, in most cases, already in place.

12) Electricity generation is avoided over the long run.

13) Start-up costs are less than $500,000. This figure includes one full-time equivalent state government employee and eight engineers (for eight municipalities).

14) If done correctly, the RFP process guarantees that all Florida municipalities are given the opportunity to participate in the program and potentially share the benefits.

15) Similar programs are likely to succeed in other states if a comparable system for measuring energy savings is used. The engineering calculations used to measure energy savings in Florida are basic, proven, and accepted.

16) While most energy engineers will be able to recommend energy conservation measures that will save their annual salary many times over, implementation of the energy-conservation measures are often limited to the municipality's budget. Energy-saving technologies often require initial capital expenditures.

17) No program evaluation information was available for NCSL.

Future Direction/Legislative Priorities

The Florida Legislature will consider a number of energy-related initiatives in its 1993 session. The most notable energy legislation to emerge in 1992 came out of the Committee on Regulated Services and Technology. This proposed legislation establishes a residential energy-rating system for the state, promotes solar and other renewable energy technologies, promotes integrated resource planning and demand-side management, extends thermal effeciency standards to addtional products, and strengthens existing building construction codes.

Florida's counties and cities are prohibited from restricting construction of solar energy devices for residences. In the summer of 1991, a Florida court ruled in favor of a homeowners' association that had sought to restrict the construction of a residential solar device. Legislation prohibiting homeowners' associations from restricting the construction of residential solar devices passed the Legislature in 1992.

The Florida Legislature is also likely to consider alternative fuel, waste-to-energy, new IRP, and cogeneration legislation in 1993 (see chapter note 3).

Chapter Notes

1   Florida Governor's Energy Office, *1989 Annual Report to the Legislature* (December 1989): 31.

2   Danny S. Parker, *Supplying Electricity Through Greater Efficiency: A Research Proposal for a Model Residential Energy-Efficiency System for the State of Florida* (January 1990): 1.

3   Florida Department of Community Affairs (FDCA), *1991 Agency Functional Plan*, 35.

4   Florida Energy Office (FEO), *Florida's Coordinated Energy Plan 1991: July 1, 1991–June 30, 1992*, 3.

5   FEO, *GEO Annual Report*, 1989.

6   *Florida's Coordinated Energy Plan 1991*, x.

7   Florida's *1989 Annual Report to the Legislature*, 6.

8   Helen Gonzalez, National Consumer Law Center, January 1992: personal communication.

9   *Energy Plan 1991*, 3.

10   Barney L. and Lynne C. Capehart, "The History and Impact of the Florida Energy Efficiency and Conservation Act of 1980," *Florida Scientist* 50, 4, (Autumn 1987): 193.

11   *1989 Report to the Legislature*, 5.

12   *Energy Plan 1991*, 3.

13   *1989 Report to the Legislature*, 15.

14   *Energy Plan 1991*, x.

15   Ibid., 3.

16   *1989 Report to the Legislature*, 22.

17   *Energy Plan 1991*, vi.

18   John Blackburn and Barney L. Capehart, *Economic Analysis of the Florida Energy Conservation Standards Act*, 1989, 3.

19   Ibid., 1.

20   Ibid., 7.

21   FEO, *Fact Sheet on Florida Solar Users Network*, November 1991.

22   FEO, *Fact Sheet on Mass Transit Demonstration Project*, November 1991.

23   Danny Porter, Florida Solar Energy Center, March 1991: personal communication.

24   Richard Shine, Florida PSC, November 1991: personal communication.

25   Ibid.

26   Ibid.

27   Pursuant to Chapter 186.801, Florida Statutes, Florida's electric utilities submit "ten-year plans" that identify power generating needs and the location of proposed power plant sites. This information was derived from the most recent plan.

28   *1989 Report to the Legislature*, 48.

29   Ibid., 47.

30   *Energy Plan 1991*, 2.

31   Tom Batchelor, Florida House Committee on Regulated Services and Technology, January 1992: personal communication.

32   Clarence Council, WAPA, October 1991: personal communication.

IOWA

I owa's dependence on imported energy affects its interest in energy efficiency and the use of renewable resources. Iowa relies on fossil fuels for 95 percent of its energy needs and imports more than 98 percent of its total energy from sources outside the state.[1] Iowa is an agricultural state with relatively few indigenous resources for energy production. Until sustainable agriculture was introduced in the twentieth century, the Iowa agricultural sector depended almost exclusively on fossil fuel. Petroleum is Iowa's main energy source (**figure 31**). Coal and natural gas are the second and third largest sources of energy, respectively. Nuclear power represents a small percentage—3.6 percent in 1988—of Iowa's total energy consumption. Renewable energy sources—solar, wind, wood, agriculture crops, solid waste, and hydroelectric power—currently supply less than 1 percent of the state's energy, with corn-derived ethanol and hydroelectric power accounting for the majority. Iowa's reliance on fossil fuels is mirrored by the other five states detailed in this study.

Because Iowa is known for energy programs that apply to all traditional energy-consumption sectors, legislators attempt to address all sectors, especially transportation. Many states largely ignore the transportation sector and attempt to reduce energy consumption in the other three. Oil price fluctuations have significant effects on Iowa's oil-intensive transportation sector. A 10-percent rise in the price of oil is said to drain $500 million from Iowa's economy.[2]

Figure 31
Iowa Energy Use by Source, 1988 (942 Total Trillion Btus)

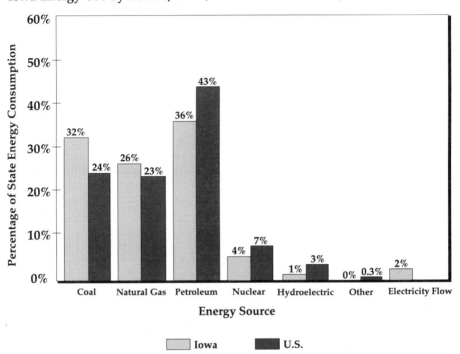

Source: Energy Information Administration, State Energy Data Report, Consumption Estimates, 1960–1988, DOE/EIA-0219, 1988.

Iowa is active in both energy efficiency and environmental protection. In recognition of the close interrelationship of energy resources, the environment, and the economy, in 1990 the State of Iowa developed the innovative Integrated Program for Energy Conservation and Environmental Protection. This program combines the Energy Efficiency Act of 1990 (Senate File 2403) and the Iowa Groundwater Act of 1987, and has three components to address energy consumption—the Waste-to-Energy, Solid Waste Management Program; the Agricultural-Energy Management Program; and the Energy Resources Development Program. Waste-to-energy programs have cost-shared regional and community demonstration projects in waste reduction, recycling, and energy recovery from refuse-derived fuel (RDF) in nearly half of Iowa's counties. The agricultural programs have developed integrated farm management demonstrations to show that reduced chemical use enhances the profitability of crop production. The energy-resource area promotes biomass energy and automated techniques for energy resource assessment and planning. All programs are designed to help Iowa residents make the transition to energy-saving, pollution-preventing practices. The Integrated Program for Energy Conservation and Environmental Protection framework is transferable to other states. The program was funded originally over five years with $18 million in stripper well oil-overcharge funds.[3]

Iowa also is known for its comprehensive energy-efficiency bill of 1990, Senate File 2403, which addresses energy consumption in all of Iowa's energy sectors. With the bill, a number of energy initiatives were started, including a building energy-efficiency rating system, demonstration grants for alternative fuels, a new Iowa Energy Center, IRP by utilities, and a requirement that all of Iowa's investor-owned utilities (IOU) spend at least 2 percent of operating revenues on energy-efficiency programs, a global warming study, a comprehensive statewide transportation study, car care clinics (for better efficiency), and a telecommuting study. The successful focus group-advisory group process used to develop Senate File 2403 is worthy of mention. Iowa energy staff spent over a year organizing data and identifying the stakeholders who would participate in the energy-efficiency strategy. Appendix 5 provides a brief summary of Iowa's Comprehensive Energy Plan. The focus group-advisory group process and the entire 1990 comprehensive energy plan is described in *Environment, Economy, Energy: Iowa at the Crossroads*, available through Iowa's Department of Natural Resources (DNR).

Iowa's total energy consumption (**figure 32**), has varied measurably in the last 20 years. After a peak in 1979, energy consumption steadily declined until 1987; a 27-percent reduction in petroleum use and a 30-percent reduction in natural gas use between 1979 and 1987 were responsible for this drop in consumption. The trend in decreased petroleum and natural gas use, however, has been partially offset by a 31-percent increase in coal use from 1979 to 1987, primarily for the production of baseload electricity.[4] The increase in coal production has drawn attention to the associated increase in air emissions.

The 1988 data show a reversal in the consumption trend. Iowa's 1988 total energy consumption returned to the 1980–1981 level, an increase of more than 7 percent from 1987. A marked growth in coal, petroleum, and natural gas use occurred during this period, due to inexpensive energy prices. If more energy-efficiency measures are not aggressively pursued, overall energy consumption in Iowa is projected to increase at an average of 1 percent per year through 2010; this rise is attributed to the forecasted expansion in Iowa's economy.[5]

Iowa has brought the private sector into its energy-efficiency programs with profit opportunities. Private sector companies now identify, design, and finance energy improvements in state agencies. In 1985, the Iowa legislature changed the law (Senate File 303) to allow state agencies to enter into lease-purchase agreements for energy-management improvements. To address the energy-efficiency needs of state facilities, the State of Iowa

Figure 32

Total Annual Energy Consumption, Iowa, 1970–1988

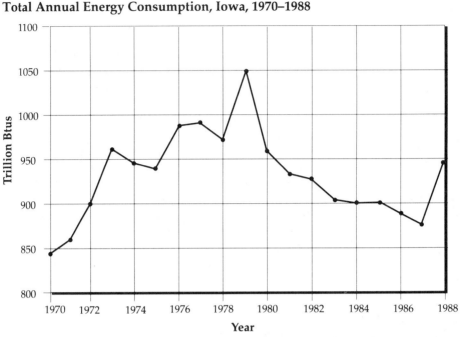

Source: Energy Information Administration, State Energy Data Report, Consumption Estimates, 1960–1988, DOE/EIA-0219, 1988.

Facilities Improvement Corporation (SIFIC) was formed to coordinate efforts of state agencies to identify and implement cost-effective, energy-management improvements, using financial and technical teams. This was the first step in Iowa's comprehensive building energy-management program for public and nonprofit facilities.

Iowa officials project that approximately $300 million will be needed to make all cost-effective, energy-management improvements in schools, hospitals, local and state government facilities, and nonprofit agencies. Iowa defines as "cost-effective" those improvements that pay for themselves in aggregate with energy savings in six years or less. Accomplishing that goal requires two phases: *identify* energy waste and *install* needed improvements. By the end of FY1991, Iowa had completed comprehensive engineering analyses that identified $97 million in needed improvements. Approximately $32 million in improvements were installed by the end of FY1991.

To address the needs of schools, private colleges, hospitals, cities, and counties, separate "energy bank" programs were established. Zero-interest and low-interest loans, as well as technical, legal, and financial consulting services were made available to participating jurisdictions to identify and implement cost-effective, energy-management improvements. The energy bank program consultants arrange for energy audits, comprehensive engineering analyses, and financing. While some facets of the program are similar to SIFIC, the Iowa School Energy Bank program offers financing for projects ranging from $15,000 to $10 million, available from a regional lender rather than from bond sales like the SIFIC.

Energy Consumption by Sector

In 1988, approximately 942 trillion Btus of energy were consumed in Iowa. The breakdown of this use by residential, commercial, industrial, and transportation sectors is illustrated in appendix 3, where Iowa's energy sector use is compared with the six other states.

The Industrial Sector

The industrial sector is the largest energy consumer in the Iowa economy, accounting for 37 percent (**figure 27,** page 135) of total energy consumption in 1988; agricultural production uses approximately 27 percent of this total. Crop and livestock production account for about 10 percent of Iowa's industrial energy use. Other activities in this sector include manufacturing, construction, mining, and agricultural operations.

Nearly 43 percent of the total natural gas demand in 1987—the largest of any sector—came from industry, where natural gas is used for space and water heating, to power manufacturing processes, and as a raw material (primarily for organic fertilizers). Industry also leads in coal use and is second in electricity use. Petroleum use in this sector occurs mainly for a variety of off-highway agricultural and construction industries.

During the next 20 years, the total energy requirements for the industrial sector are predicted to increase slowly, at less than 1 percent per year.[6] An increase in industrial activities is expected to be offset by more efficient industrial facilities combined with a shift toward the production of goods that require less energy per dollar of final product.

The Transportation Sector

The transportation sector is the second largest consumer of energy in Iowa and accounts for 25 percent of the state's total energy use (**figure 28,** page 136). As in many states, the transportation sector uses more petroleum than any other sector.

Gasoline and diesel fuel are the primary forms of energy consumed by this sector. Currently, more than 2.7 million vehicles are registered in Iowa (including 1.8 million automobiles). These motor vehicles are responsible for approximately 94 percent of the gasoline and 55 percent of the diesel fuel

166

burned in Iowa.[7] Iowa vehicles average only 12.4 miles per gallon (mpg). This compares to the national average of 14.2 mpg, with state-estimated vehicle fuel efficiencies ranging from 10.8 mpg to 19 mpg. The poor mpg average may be due to a relatively large number of older, less efficient cars driven in Iowa.

Experts predict that petroleum consumption by the Iowa transportation sector will increase less than 1 percent per year for the next 20 years. Improvements in vehicle efficiencies are expected to increase the state's mpg average in the future. These efficiency gains will be offset by a projected 18 percent increase from 1987 to 2010 in the number of vehicle miles traveled.

The Residential Sector

The residential sector consumes about 24 percent of Iowa's total energy demand (**figure 29,** page 137). Residences consume the highest amount of electricity for any Iowa sector and the second highest amount of natural gas at 38 percent and 32 percent, respectively. Residential petroleum consumption is significant.

The increase in home energy efficiency keeps pace with the increase in number of new households. Since 1972, the average efficiency of electrical appliances has risen dramatically. The efficiency of new refrigerators has risen by 80 percent, freezers by 67 percent, and washing machines by more than 50 percent. Natural gas use per residence has also declined due to improvements in the efficiency of natural gas furnaces and water heaters coupled to increased levels of home weatherization. Fuel oil consumption continues to decrease as it is replaced by natural gas.

Total energy demand in Iowa's residential sector is projected to increase slightly (less than 0.5 percent per year) for the next 20 years. Efficiency gains are expected to balance the increase in the number of residences.

The Commercial Sector

The commercial sector is the least energy consuming of the four Iowa sectors outlined; it accounted for only 15 percent of Iowa's total energy consumption in 1988 (**figure 30,** page 138). All nonmanufacturing business establishments are represented by this sector, including motels; restaurants; wholesale businesses; retail stores; health, social, and educational institutions; and federal, state, and local governments. Natural gas, used mainly for space heating, is the most common fuel in this sector.

The Iowa economy is predicted to shift steadily toward a more service-oriented economy, and the commercial sector could encounter the most change in the next 20 years. Consistent with the national trend, employment in service-related companies is expected to increase 25 percent by 2010, much greater than the average 4-percent growth predicted for all sectors.

Increased activity and floor space in this sector will increase the amount of space heating needed. Without energy-efficiency measures, the total energy consumption in this sector is projected to increase an average of 2 percent per year over the next 20 years.

Energy-efficiency Statutes

Code Chapter 19:

   19.34 Energy conservation lease-purchase. A state agency may lease real and personal properties and facilities for use as, or in connection with, energy-conservation measures that pay for themselves in energy savings in six years.

Code Chapter 72:

   72.5 Life-cycle cost.

   1) A contract for public improvement or construction or renovation of a publicly owned building shall not be let without preparation of a design proposal reflecting the lowest life-cycle cost possible in light of current technology, and a cost-benefit analysis of any deviations from the lowest life-cycle cost analysis.

   2) DNR, Department of Management, state building code commissioner, and state fire marshal shall develop standards for a statewide building energy-efficiency rating system.

Code Chapter 93:

   93.12 Implementation of energy conservation measures—state board of regents. Mandates the board of regents to perform comprehensive engineering analyses of its buildings and implement the improvements identified in the analyses. The cost may be reduced by using funds borrowed from the SIFIC. Requires the board to annually report to the DNR its progress.

   93.13A Energy conservation measures identified and implemented. Requires state agencies, political subdivisions, schools, area education agencies (AEAs), and community colleges to identify and implement all cost-effective, energy-management improvements for which financing is made available by the DNR to the entity.

   93.19 Energy bank program. Establishes an energy bank program to disburse oil-overcharge funds for energy audits, to provide loans and technical assistance, and to provide self-liquidating financing for implementation of energy-conservation improvements.

   93.20 Energy loan fund. Establishes an energy loan fund to make loans to school districts, AEAs, and community colleges for energy-conservation measures that have a six-year simple payback. Allows school districts, AEAs, and community colleges to enter into lease-purchase financing with DNR or its appointed representatives.

Requires schools, cities, and counties to design and construct the most energy cost-effective facilities feasible while using financing made available by the department to cover incremental costs above the minimum building code energy-efficiency requirements, unless other lower cost financing is available.

93.20A Self-liquidating financing. Allows the DNR to enter into financing agreements with school districts, AEAs, community colleges, and cities or counties to provide financing for implementation of energy-conservation facilities.

Code Chapter 470:

470.2 Policy—analysis required. The General Assembly declares energy managementof primary importance in the design of publicly owned facilities. Life-cycle cost analyses shall be conducted in the design phase of such purchases.

Code Chapter 476:

476.1A Applicability of authority—certain electric utilities. Small electric utilities and cooperatives are subject to fee assessment to support the Iowa Energy Center and Center for Global Warming and must file efficiency plans. The IUB may permit joint plans to be filed, or waive the requirement if the utility demonstrates that existing programs are highly energy efficient.

476.1B Applicability of authority—municipally owned utilities. Municipally owned utilities are subject to regulation encouraging alternate energy production and are subject to assessments of fees to support the Iowa Energy Center and Center for Global Warming. They must also file efficiency plans.

476.6 (17) (89 Code Supplement). Beginning January 1, 1992, no rate increases shall be approved for a rate-regulated utility unless the utility has a comprehensive energy-management program in effect. Energy-efficiency programs shall be considered capital items for rate-making purposes, but will not be allowed as a separate cost or expense in customer billing. The board shall allow utilities to recover the costs of energy-efficiency programs that are cost-effective.

476.8 Utility gas charges and service. Every public utility must furnish programs encouraging energy efficiency and renewable energy resources.

476.10 Investigations—expense—appropriation. The board and the office of consumer advocate may add personnel to review and evaluate energy-efficiency plans and implementation of programs.

476.10A Funding for Iowa Energy Center and Global Warming Center. All gas and electric utilities shall fund the centers with one-tenth of total annual gross operating revenue from intrastate public utility operations.

476.17 Peak load energy conservation. The board may promulgate rules requiring public utilities to establish reduction of consumption at peak loads.

476.19 Rate-regulated utility energy-efficiency plans and budgets shall be designed to spend annually the following percentage of gross operating revenues: electric—2 percent, gas—1.5 percent. Also makes provisions for cost recovery and plan financing.

476.41 Alternate energy-production facilities: Purpose. Defines state policy to encourage development of alternate energy-production facilities.

476.62 Energy-efficient lighting required. All publicly owned exterior lighting shall be replaced when worn out exclusively by high-pressure sodium lighting or equivalent with better energy efficiency.

476.63 Energy-efficiency programs. The division shall consult with the DNR to develop and implement public utility energy-efficiency programs.

Code Chapter 93:

93.7 Subsection 1, Code 1987 Requires that a comprehensive energy plan be developed and submitted to the General Assembly by January 15, 1990, and that it be updated biennially.

93.20C Requires the Iowa DOT to use the SIFIC to conduct technical engineering analyses on all facilities and to implement energy-management improvements with aggregated six-year payback. (SIFIC is discussed under "Energy Efficiency Programs and Initiatives.")

HF 2469 (1989, 72nd General Assembly, 2nd session). Requires the division of community action agencies in the Department of Human Rights, in cooperation with the DNR and Utilities Board, to conduct a two-year pilot project to determine the most economical and effective means of maintaining low-income Iowans' access to heating fuels and to develop more effective programs for weatherizing residences and achieving energy conservation.

18.115, Subsection 4, Code Supplement 1989, amended. Requires that the state vehicle dispatcher, effective January 1, 1993, implement a system of uniform standards to assign vehicles available for use to maximize the average passenger miles per gallon of motor vehicle fuel consumed. Once rules are adopted, this requirement will apply to the department of transportation, the department of the blind and to the board of regents.

18.115, Subsection 4. Also requires that the state vehicle dispatcher, and any other state agency or local governmental political subdivision purchasing new motor vehicles for other than law enforcement purposes, shall each year purchase new passenger vehicles and light trucks such that the average fuel efficiency for the fleet of new passenger vehicles and light trucks equals or exceeds the corporate average fuel economy (CAFE) standard for the vehicles' model year as established by the U.S. Secretary of Transportation.

Chapter 455B.263(1), 455B.480-490, 455E, Groundwater Protection Act. Provides for a plan to protect Iowa's groundwater resources. Included in the strategies are the Leopold Center for Sustainable Agriculture, the Integrated Farm Management Demonstration Program, educational programs, the Center for Health Effects of Environmental Contamination, solid waste comprehensive planning, watershed protection programs, and the groundwater protection funds. (These programs encompass energy efficiency as a central theme.)

Chapter 476.41 Provides for the establishment of a buy-back rate for promotion of power sold to utilities by small energy producers (alternative energy producers).

Chapter 185C Establishes the Iowa Corn Promotion Board and its ethanol responsibilities.

Chapter 422.45 (11) Ethanol Tax Incentive provides a $.01/gallon tax incentive for 10 percent ethanol blended fuels.

Chapter 455C Requires a $.05/bottle deposit to encourage recycling.

Energy-efficiency Programs and Initiatives

Building Energy Management

While many Iowa state agencies have traditionally practiced sound energy management, lack of capital prevented installation of many energy-saving improvements. To address this problem, 1985 legislation (Senate File 303) authorizes state agencies to use lease-purchase financing to make energy-management improvements. Constitutionally, general obligation bonding by the state requires voter approval by referendum. Lease-purchase financing does not require a referendum and enables state agencies to pursue energy-conservation measures, thus avoiding the lengthy referendum process.

In addition to authorizing lease-purchase financing for energy management, the 1985 legislation requires that energy-conservation projects have an average payback of six years or less. This law establishes a minimal, nonspecific framework for energy management but does not create the nonprofit corporation ultimately formed to implement it. The 1985 law is key to the implementation of the DNR Building Energy Management Program Objective: To install all energy-management improvements with an aggregate payback of six years or less in all public and nonprofit facilities by 1995. To meet this objective, the DNR established a new revenue stream ($10 million annually) to finance all conservation measures.

As described previously, an estimated investment of $300 million for capital improvements will be required to meet the Building Energy Management Program Objective and will result in annual energy savings between $50 to $60 million when this goal is achieved.

The State of Iowa Facilities Improvement Corporation

To promote the use of lease-purchase financing for energy management for state agencies, a financial team recommended the creation of a nonprofit corporation. The SIFIC was organized to provide energy-related capital improvements for state agencies by financing leveraged energy savings. A technical team works concurrently with a financial team (**figure 33**). The technical group develops a system by which energy-conservation measures are identified and selected, and the financial team organizes the financing needed to install those improvements.

Initial work resulted in the current process by which energy improvements are accomplished. Agencies that participate in the lease-purchase program have an energy audit, followed by an engineering analysis of their facilities. Audits are provided free by a firm under contract to the Iowa DNR. Engineering analyses are conducted by a consultant under contract to the Improvement Corporation. Costs for engineering analyses are incorporated into the lease payments. The Improvement Corporation staff reviews the engineering consultant's recommendations and works with the agencies to select specific management improvements for design and implementation. Following detail design by the engineering consultant, contractors bid on installation of the selected management improvements. The contractor chosen must be able to guarantee that the proposed improvement(s) will save a specified amount of energy, or that the value of the savings will be sufficient to cover the lease payments.

Bonds are issued by the Improvement Corporation to obtain the capital necessary to install improvements with an average payback of six years or less. After the measures are installed, energy savings are monitored for performance and the savings are used to repay the lease. If savings fall short of the projected amount, the contractor makes improvements to correct the situation or the agency is reimbursed by the contractor, according to any insurance or guarantee.

Iowa's comprehensive state and local government building energy-management initiative reduces public sector energy consumption, thereby reducing the burden of escalating energy costs on taxpayers and stemming the outflow of dollars from the Iowa economy. In addition, the installation of energy-management improvements is a labor intensive activity that provides employment to Iowans and stimulates the energy-management services industry.

172

Figure 33
State of Iowa Facilities Improvement Corporation Process

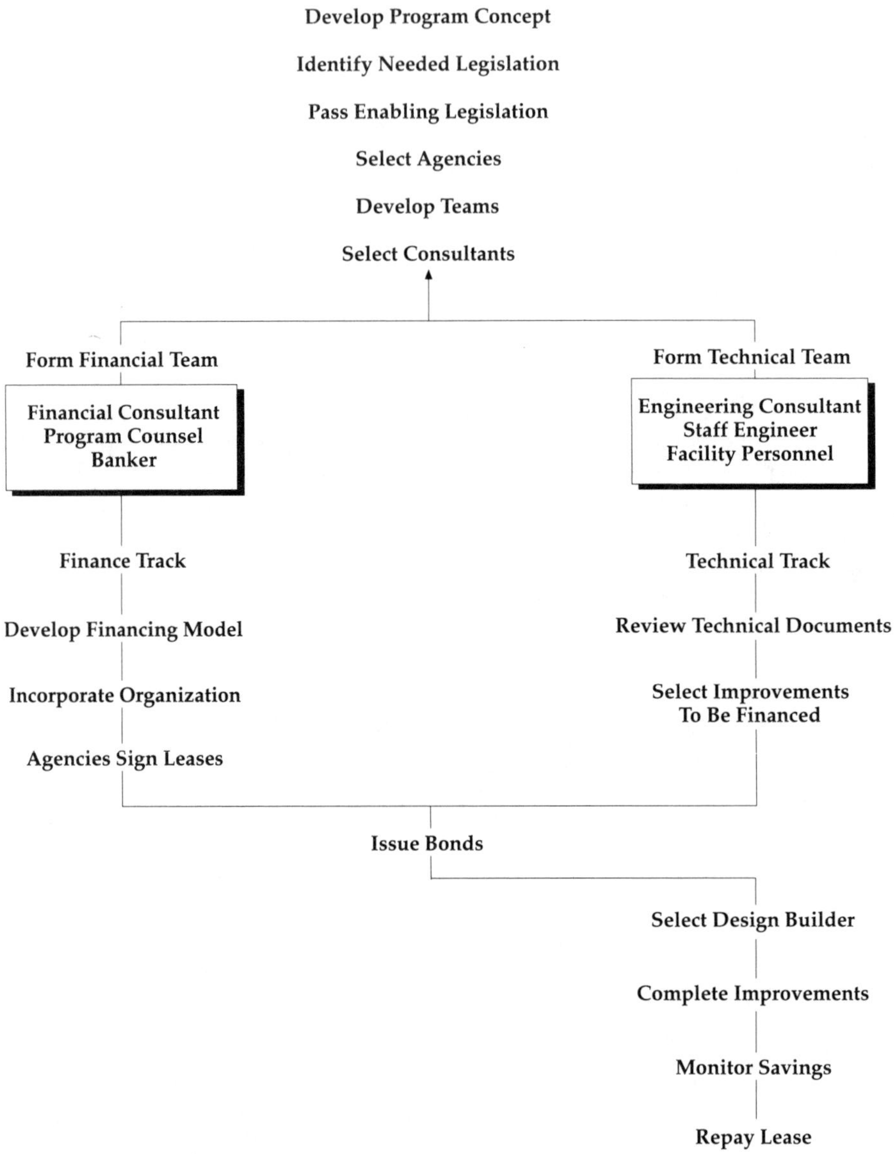

Develop Program Concept

Identify Needed Legislation

Pass Enabling Legislation

Select Agencies

Develop Teams

Select Consultants

Form Financial Team   Form Technical Team

Financial Consultant   **Engineering Consultant**
Program Counsel   **Staff Engineer**
Banker   **Facility Personnel**

Finance Track   Technical Track

Develop Financing Model   Review Technical Documents

Incorporate Organization   Select Improvements
                                                 To Be Financed

Agencies Sign Leases

Issue Bonds

Select Design Builder

Complete Improvements

Monitor Savings

Repay Lease

Source: Energy Management Financing for State Facilities and Public Schools, Iowa Department of Natural Resources, 8, 1988.

Program Questions and Answers

What is an energy audit?

An energy audit is a study of facilities; fuel bills; energy use; cost data; occupancy rates; heating, cooling, and electrical equipment; and operation and maintenance procedures. Auditors interpret these data to determine whether energy-management improvements will save money or whether further steps, such as an engineering analysis, are needed. Participating schools receive a copy of an energy-audit report that details the study results.

What is an engineering analysis?

An engineering analysis is a more detailed study of the building exterior, HVAC equipment, and other energy-using systems. It includes simulations of building energy use and identifies opportunities for energy management. The analysis provides cost estimates for improvements and projections of how much money and energy could be saved annually as a result of these improvements. It also includes an estimated payback time for identified improvements. An analysis is needed for buildings with complex HVAC systems.

What energy-management improvements qualify for financing?

Improvements with an average payback of six years or less are financed. Examples of specific improvements include lights and ballast replacement, boiler replacement and controls, time clock and set-back controls, insulation, weatherstripping, improvements to air handler systems, and air destratification fans.

The program is designed to be cash-flow neutral for state government. The lease rental payments are designed to be sufficient to retire bond principal and interest, which have been structured to be paid with energy savings from the installed improvements. Thus, the lessees—Iowa state agencies in this case—will have no additional budget effects from financing.

If the energy savings are actually greater than the conservative projection on which the financing was based, or if energy prices are greater than projected, the budget effect will be positive for the state agencies even before the debt is retired. Cost savings to the state will, of course, balloon when the debt is repaid. Energy cost projections to provide the basis for future savings begin with an initial cost of $2.75 per million Btus for natural gas and the current price of electricity at each location, inflating at 4 percent per year.

The first Iowa state agencies targeted by the SIFIC for energy improvements were general services, human services, and the department of corrections. In September 1986, the state issued $12 million in Series A bonds to help these state agencies make energy improvements. The SIFIC program was extended with Series B bonds, worth $30 million in the late 1980s, to the Iowa department of transportation, department of the blind, the department of public safety, and the state university system. Other agencies use their own funding to make energy improvements. These self-financed groups include the board of regents, the fair board, the department of cultural affairs, and the DNR.

The DNR is expected to track a total of $56 million in energy improvements in state facilities, including the state board of regents facilities, resulting in over $9 million in energy savings annually by its completion (**table 24**). As of June 1991, all $56 million in improvements were identified. More than $18 million of these improvements—32 percent of the total goal—have been completed, saving the state an estimated $3 million annually in energy costs and resulting in a simple payback of six years.[8]

Table 24
Iowa Energy Program Savings

Program	Responsible Group (s)	Improvement Cost	Completed Improvement Cost	Estimated Annual Energy Savings on Completed Improvements	Estimated Payback on Completed Projects
Building Energy Management (1985-1991)	State Agencies (SIFIC)	$56,000,000	$18,115,000	$3,009,000	6.0 years
	Schools (School Energy Bank)	$70,000,000	$13,491,000	$2,249,000	6.0 years
	Hospitals (Hospital Energy Bank)	$70,000,000	$289,000	$48,000	6.0 years
	Local Gov't (Local Gov't Energy Bank)	$50,000,000	$76,000	$13,000	5.8 years
	Nonprofit Organizations	$70,000,000	$289,000	$48,000	6.0 years
Motor Vehicle Fuel Reduction Program (1987-1989)	19 cities	$3,000,000	$2,570, 124	$4,006,042	7.7 months

Source: Iowa Department of Natural Resources, 1991.

The Iowa School Energy Bank (House File 2387)

A second part of Iowa's Building Energy Management Program was to institute a state-administered program to finance energy improvements in the 430 Iowa public school districts, area education agencies, and community colleges. In 1986, legislation was passed (House File 2387) to require all Iowa school facilities to conduct energy audits once every five years and to authorize the development of an energy bank for schools. Under this direction, the Iowa School Energy Bank was established. To facilitate development of the School Energy Bank, the 1986 enabling legislation was revised in 1987 to authorize schools to enter into full-term leases that bind the district beyond the current fiscal year. In addition, the bank's constituency was expanded to city and county governments.

The three-phase program includes energy audits, engineering analyses, and financing for improvements. Energy audits are provided free by an auditing firm on contract to the DNR. Six-month, interest-free loans are offered to the schools for engineering analyses, which evaluate potential energy improvements suggested in the energy audit. The school district then selects the recommended improvements it wants to implement. The average cost of all improvements to all facilities in the school district must have an average simple payback of six years or less.

Lease financing is offered for improvements with a regional bank at a group municipal financing rate. The program offers a flexible payment plan where the school districts may select the length and payment schedule of the lease. Lease rental payments may be designed to be equal to the expected energy savings and thus be cash-flow neutral. Alternatively, a district may choose to lengthen the lease term to achieve a positive cash flow or to shorten the term to pay off the improvements more rapidly. When the energy improvements are implemented and the lease payments are completed, school districts will realize additional funds for educational priorities.

In 1989, the governor and the General Assembly reinforced their commitment to energy efficiency in Iowa by passing Senate File 419. This legislation requires every Iowa local government and public school district to identify and implement all energy-conservation measures for which financing is available through the DNR.

Public schools are predicted to install $70 million in energy-efficiency improvements, as indicated in **table 24**. As of June 1991, $41 million in energy improvements were identified through energy audits and $13 million of these improvements have been installed. Thus, over 19 percent of the total improvements anticipated by the Building Energy Management Program are completed.[9]

The Hospital Energy Bank

Additional energy-management programs were developed following the 1985–1987 legislative authorization of lease-purchase financing for energy management. In 1988, the DNR contracted with the Iowa Hospital Association to provide a comprehensive program for the financing and acquisition of energy-management improvements for association member hospitals. The Hospital Energy Bank uses a master lease similar to that of the School Energy Bank and offers financing based on energy audits. It also offers credit enhancement to enable participation by hospitals with less favorable credit ratings. Of an anticipated $70 million in expected improvements, $750,000 of these have been completed through the Hospital Energy Bank.[10] Total energy savings from these completed improvements are estimated to have a payback of six years.

Nonprofit and Local Government Energy Bank Savings

Also in 1988, United Way of Iowa initiated a comprehensive energy-saving program for nonprofit organizations. In 1990, 11 nonprofit agencies completed technical engineering analyses that identified improvements totaling $104,000. Selected energy improvements are financed through the Iowa Power Company, with the DNR offering grants to pay for up to 20 percent of the total costs.

The latest efforts include expansion of energy-management improvement programs to local governments. In 1990, the Building Energy Management Program began to help cities, counties, and municipal utilities identify and finance cost-effective energy-management improvements. Local government facilities include buildings, street lighting, waste and waste water treatment plants, and power plants. The program goal is to invest $50 million in energy improvements by 1995. As of June 1991, energy improvements costing $76,000 have been installed to provide an estimated $13,000 in annual energy savings.[11] Private colleges, private schools, and other nonprofit organizations are served in separate programs that are similar to the other energy bank programs.

The Motor Vehicle Fuel-reduction Program

In 1987, Iowa began a traffic-signal improvement demonstration program—the Iowa Motor Vehicle Fuel Reduction Program—similar to traffic-synchronization programs in California, Florida, Illinois, Maryland, Michigan, Missouri, New York, North Carolina, Pennsylvania, and Virginia. Original funding came in 1986, when the General Assembly authorized $3 million in Exxon funds as part of Iowa's State Energy Conservation Program. The DNR administers oil-overcharge funds in the state and, by cooperative agreement, the Iowa DOT is authorized to administer the program.

The payback on Iowa's program has been exceptional—less than eight months.[12] The objective of the program was to improve the efficiency (timing) of traffic signals, thereby reducing fuel consumption, eliminating delays in travel time, decreasing automobile pollution, and improving traffic safety.

A 1989 report states that 299 intersection improvements were made in 19 Iowa cities evenly distributed in size—small cities (less than 10,000 population), medium cities (between 10,000 and 50,000 population), and large cities (more than 50,000) were equally represented in the program.[13] Improvements in traffic signal efficiencies included upgrading isolated intersections, coordinating arterials with time-based coordinators, and upgrading the control of downtown network systems. Modern traffic control equipment, including microprocessor controllers and solid-state controls, were installed in many cities, replacing traditional traffic controllers that use electric motors, relays, and switches. Because of a lack of large, central business districts in Iowa, there was little need to install complex network systems. This fact makes Iowa's energy-saving figures even more impressive, since the savings were achieved in what many consider rural areas.

In addition to the signal equipment and signal-timing improvements, the Iowa program included three components:[14]

1) A traffic engineering consultant conducted before and after studies and economic evaluations of each city's program.

2) Iowa initiated a publicity campaign to improve public awareness of the energy savings associated with traffic light synchronization, something similar programs in other states did not do. The "Iowa Signals Go" campaign included press conferences, press packets, radio and television public service messages, a promotional videotape, and newsletter articles.

3) A technology transfer program was conducted for technical staff and local and state decisionmakers. Most states only target traffic engineers as part of their traffic light synchronization programs.

It is estimated that more than $4 million in energy savings will be realized annually with an average simple payback of less than eight months for all the improvements, as shown in **table 24**. It is estimated that the changes in traffic signals will save almost 280,000 gallons of fuel ($279,850) each year. Other estimated annual benefits include a savings of 845,800 hours in travel time ($2,833,430) and the avoidance of delays and almost 150 million stops ($892,762).[15]

Other benefits of similar programs include improved traffic safety, improved equipment reliability, reduced maintenance costs, improved pedestrian safety, and improved non-traffic-related safety through emergency vehicle pre-emption.[16]

Municipal Utility Community Energy-saving Program

A recent public television series voted Osage, Iowa, the "Conservation Capital of the Nation." A comprehensive energy-conservation program was implemented in 1974 that saves the city's 3,800 residents $1.2 million annually in avoided energy costs. With the program, the city avoided construction of a new electrical generating facility in 1984, and has been able to delay construction of any new generating facility until after 2000.[17]

The program involves 95 percent of the utility's customers, and is directed at the inefficient use of electricity and natural gas. Program activities include weatherizing homes, providing insulating wraps for hot water heaters and pipes, offering aerial and ground-level thermogram checks for heat loss, giving away energy-saving compact fluorescent bulbs for interior use, making available high-pressure sodium bulbs for street lights, publishing a newsletter, sponsoring an educational outreach program, providing free energy audits to industry, and planting trees around buildings to reduce exposure to the sun and save on cooling bills.

Organized and led by the Osage municipal utility, the program is responsible for avoiding the need to burn 150 tons of coal, bringing 100 new jobs to Osage, and increasing the economic vitality of the area.[18] Other municipal utilities in Iowa have followed the lead of Osage to bring greater efficiency to their customers.

Integrated Resource Planning

Most states have only utility commission orders to guide their IRP process. Iowa enters the implementation phase of its IRP process—unique because the state formally codified its IRP directives in 1990 and 1991—after an extensive development process.

In February 1992, all IOUs in Iowa will implement energy-efficiency programs in their respective territories. Municipal and rural electric cooperatives are not required to invest a portion of their gross revenues in efficiency investments, but instead must submit efficiency plans to the Iowa Utilities Board (Iowa's public service commission) for review.

In 1977, legislation (chapter 476A) required utility companies to obtain "construction certificates" to prove the utility had considered all "cost-effective" alternative energy sources. In the early 1980s, the Utility Board mandated that utilities grant low-interest loans for investment in energy efficiency. Utility companies fought the mandate and took their case to the state supreme court, which ruled in their favor of the utility companies, saying that the Utilities Board had no right to mandate low-interest loans by utilities for energy-efficiency investments.

The General Assembly enacted Senate File 419 in 1989. This act requiring public utilities to operate in an efficient manner and authorizing the Utilities Board to adjust profit levels or revenue requirements to provide utilities with incentives for energy efficiency. The act further provides for an increase in the efficiency of public lighting.

In 1990, the General Assembly enacted Senate File 2403 (Iowa Code, Section 476), creating the Iowa Energy Center at Iowa State University to conduct research and development in energy efficiency and alternative and renewable energy-production systems. The act further requires electric utilities to develop comprehensive energy-management programs. The associated regulations to Senate File 2403 require that program plans include forecasts, an assessment of supply options and avoided costs, options screening, and implementation plans for the selected programs. The Utilities Board representatives worked for the passage of Senate File 2403, and were enthusiastic about legislative interest in DSM instead of the traditional emphasis on supply-side strategies to meet new electrical demand.[19] Appendix 4 contains Iowa's 10 goals associated with Senate File 2403.

Senate File 2403 requires utilities to spend a minimum of 2 percent of gross operating revenues for electric rate-regulated utilities (1.5 percent for gas-regulated utilities) on energy-efficiency programs. A cost-effectiveness test is applied. The Utilities Board adopted rules for energy-efficiency plans on March 15, 1991, and adopted cost recovery rules on April 10, 1991.

Senate File 2403 allows ratebasing and cost recovery of DSM outside conventional rate cases. The board may treat utility expenditures and related costs of an approved utility energy-efficiency plan as capital items for ratemaking purposes. The board may also assess rewards or penalties to utilities based on the performance of the programs.

Utilities conducted previous energy-efficiency pilot programs. Iowa Power and Interstate Power have loan and rebate programs under way now. Iowa Power is conducting a project that provides direct control of air conditioning as a load management program. The program currently has 500 participants, with an additional 3,000 expected next year and an increase of 2,000 per year thereafter. Iowa Power also conducted a successful commercial lighting program in 1990.

The Iowa Energy Efficiency Act of 1990 (Senate File 2403) contained numerous utility-related codes. The following codes are most relevant to IRP.

476.1A, 1B, 1C Non-rate regulated utilities (municipal utilities, electric public utilities with less than 10,000 customers, and gas public utilities with less than 2,000 customers) shall support the Iowa Energy Center and the Center for Global Warming at the University of Iowa. Fees shall be assessed to support these centers. These utilities shall also file energy-efficiency plans with the Utilities Board. The board may permit joint plans to be filed, or waive the requirement if the utility demonstrates that existing programs are highly energy efficient.

The Utilities Board shall submit a report to the governor and General Assembly by December 1, 1992 on the level of efficiency that non-rate-regulated utilities intend to achieve, based on the above reports. By December 1, 1994, the board shall submit a report on the implementation of the plans. In both cases, the reports shall contain recommendations for legislative action.

476.6 (17) (89 Code Supplement) Beginning January 1, 1992, no rate increases shall be approved for a rate-regulated utility unless the utility has a comprehensive energy-management program in effect.

Energy-efficiency plans shall include for consideration the following programs: hot water heater blanket distribution, commercial lighting, rebate programs for purchases of energy-efficient goods (such as light bulbs), tree planting, and energy-efficiency efforts for qualified low-income residents.

Energy-efficiency programs shall be considered capital items for ratemaking purposes, but will not be allowed as a separate cost or expense in customer billings. The board shall allow utilities to recover the cost of efficiency programs that are cost-effective.

476.60 (89 Code Supplement) The board shall conduct contested case proceedings for review of energy-efficiency plans and budgets filed by rate-regulated gas and electric utilities. Energy-efficiency plans and budgets shall be designed to expend 2 percent of rate-regulated electric utility budgets, and 1.5 percent of rate-regulated gas utility budgets. This requirement shall be included in plans and budgets by January 1, 1992.

The board shall periodically require each rate-regulated utility to file a forecast of future gas or electric generating needs.

The board may also require each rate-regulated gas or electric utility to offer customer financing to install cost-effective energy-efficiency improvements in residences and businesses.

476.10A To fund Energy Center and Global Warming Center activities, all gas and electric utilities must pay a fee of 0.1 percent of total gross operating revenues derived from intrastate public utility operations. The fees are to be submitted to the state treasurer. The board must develop rules by July 1, 1991. Of the fees, 85 percent go to the Energy Center and 15 percent go to the Center for Global Warming.

476.43 (3) The board may adopt a statewide rate for purchase of energy from alternative energy-production facilities.

476.44 Utilities do not have to purchase power from alternative energy-production facilities unless the facility meets certain qualifications; utilities also are not required to purchase more than 15 megawatts from a facility at any one time.

Table 25
Iowa Matrix

IOWA Energy Program	Cost per kWh (cents/kWh)	Annual Savings ($)	Implemen- tation	Environ- mental Benefits	Start-Up Costs	Fiscal Fairness (+,-)	Probability of Success	Benefit to Cost Ratio (Benefit/ Cost)	Limitations or Special Circum- stances	Confidence
School Energy Bank	4/kWh	$16 million M	Easy (1)	Medium (2)	Medium (3)	(-) (4)	High (5)	2.61:1-E (6)	Number of school districts (7)	High (8)
State of Iowa Facilities Improve- ment Corporation (SIFIC)	4/kWh	$5.8 million E	Difficult (9)	Medium (10)	High (11)	(+) (12)	High (13)	2.61:1-E	Legislative support is crucial for bonds (14)	High (15)
Non-Profit and Govern- ment Savings Program	4/kWh	$20 million E (16)	Easy (17)	Medium (18)	Medium (19)	(-) (20)	High (21)	2.61:1-E	High-risk hospitals (22)	Low (23)
Motor Vehicle Fuel Reduction Program	Not available	$4 million E	Easy (24)	High (25)	Medium (26)	(+) (27)	High (28)	10:1-E	Rural participation (29)	High (30)

476A.2 Any electric generating facility of greater than 25 MW built after January 1, 1990, must obtain a generating certificate. Exemptions will be granted to facilities that are already well under way. Section 476A.15 allows the board to waive this requirement for facilities between 25 MW and 100 MW if the public would not be adversely affected.

All Iowa utilities, including the municipal utilities and rural electric cooperatives (RECs), are now required to submit plans to the Utilities Board detailing their energy-efficiency efforts. Although the board does not regulate the rates charged by the municipal utilities, and all but one of the RECs, the efficiency plans give the board a comprehensive overview of Iowa's energy needs.

Provisions of Senate Bill 2403 also allow alternative energy producers (AEPs) to become more competitive. They include a clarification of terminology—"qualifying facilities" as defined by federal regulations (not all qualifying facilities are "not precluded" from being AEPs); a definition of "next generating plant" as an electric utility's assumed next coal-fired baseload electric generating plant, when used to determine buyback rates; allowing a federally qualifying small hydro facility to also be a state qualifying small hydro facility; allowing the Utilities Board to set uniform, statewide buyback rates or to set individual rates for each utility; and giving the Utilities Board the authority to consider environmental factors in the determination of a buyback rate along with economic and other factors. Also, if the board adopts statewide rates, these rates will be determined on representative rather than utility-specific data; basing the buyback rate for a utility that purchases all, or substantially all, of its electricity requirements on the electric utility's current purchased power cost; and setting a limit on the amount of electricity a utility is required to buy at 15 megawatts.

Matrix Notes

The following notes correspond to the numbers in the matrix (**table 25**).

1) Implementation of this program is relatively easy due to the finite number of school districts and the limited, well-established, energy efficiency improvements necessary for school buildings.

2) Environmental benefits are medium. Additional electricity generation is avoided over the long run.

3) Start-up costs are less than $500,000.

4) Only public school districts are allowed to participate in this program.

5) Little can inhibit the success of a similar program in another state. If the program is to be effective, however, it must be supported by regional (or state) lending institutions.

6) All Iowa benefit-to-cost ratios were calculated by the Iowa DNR by multiplying the annual energy savings by the useful life of the equipment installed and dividing that total by the cost of the program.

7) School district support for this Iowa program has been strong. Larger states with more school districts might require higher start-up costs, since more improvements require more money.

8) Sufficient information was available from the Iowa DNR.

9) Implementing this program can be quite time-consuming, and setting up a new corporation and serving as the liaison between engineers, bankers, and other state agencies can be a complex task.

10) Additional electricity generation is avoided over the long run.

11) The state began with $12 million in Series A bonds to finance the program in 1986.

12) This energy program is available to state agencies. In the long run, energy dollars saved are shared between the state agency and the public. (After the loan is paid off, the savings go into the state general fund and the budget of the state agency responsible for implementing the energy-conservation measures.)

13) Probability of success is high. Responsibility for guaranteeing energy savings usually falls on the third party installing the energy-saving equipment. If the savings do not materialize, the state can be protected through terms in the lease.

14) The bond/lease process can be complicated and time consuming. It might take over a year to implement a similar program. Lessees, lawyers, bankers, and state agencies must all be aware of the program. Legislation might be needed to set up a similar corporation in other states. Legislative support is crucial for establishing the infrastructure for a similar program.

15) Sufficient material was available to review and evaluate this program.

16) This amount is contingent upon several pending leases.

17) Similar programs have been implemented with relative ease in other states.

18) Additional electricity generation is avoided over the long run.

19) Start-up costs for a statewide program of this type are less than $500,000.

20) The program audience is relatively narrow and the program benefits are not shared by all citizens. Please note that the interest rate is the same—6.3 percent in November 1991—for all cities, including those with poor bond ratings. From the cities' perspective, this is an equitable arrangement. However, for this definition of fiscal fairness we considered who benefits and who pays.

21) Similar programs have been implemented with relative ease in other states.

22) Private colleges generally are subject to higher interest rates. Hospitals can be high-risk ventures for some states. Mechanisms are in place in Iowa that allow high-risk hospitals to recoup (previously paid) higher interest rates in the form of a rebate after the hospital has made on-time payments for six years. The state then buys back the overpayments made by the hospital.

23) No formal evaluations or written documents were provided to NCSL.

24) Implementation of traffic synchronization programs is usually easy, due to a very specific, narrow audience that consists of traffic engineers in major cities. Programs like this require extra time in the planning and set-up phase, but usually do not require constant monitoring after the traffic lights are synchronized.

25) Environmental benefits in the form of reduced emissions are immediate and high.

26) Start-up costs are medium (between $100,000 and $500,000) due to the expensive traffic synchronization computer programs required. Usually, these programs require only one or two two-day meetings with traffic engineers before the computer programs are installed.

27) This program has a "+" rating, since all citizens benefit through reduced traffic delays, reduced idling time, reduced emissions, and reduced automobile gasoline bills.

28) Traffic light synchronization programs are the "stars" of state energy-savings programs in the transportation sector. Benefits are immediate and measurable.

29) Engineers need to be qualified before these traffic light synchronization programs are installed. Rural cities may not need traffic light synchronization.

30) Formal program information was available for evaluation purposes.

Future Direction/Legislative Priorities

Iowa legislators will monitor the progress of the Iowa Utilities Board as that agency implements the rules and regulations resulting from the 1990 Energy Efficiency Act (Senate File 2403). Cogeneration and waste-to-energy legislation will be considered in 1993, and there is interest in wind-energy legislation.

Iowa legislators will also monitor with interest the energy-efficiency programs of rural electric cooperatives. RECs are required to file energy-efficiency plans but, unlike the state's major IOUs, are not required to spend a percentage of gross revenues on the energy-efficiency programs. If RECs do not pursue aggressive energy-efficiency programs in the next year, legislation mandating RECs to spend a percentage of gross revenues on energy-efficiency programs is likely to emerge in 1993.[20] Iowa legislative committees will also consider consumer incentives for using alternative vehicle fuels. There is considerable legislative interest in ethanol (produced from corn).

Chapter Notes

1.   Larry Bean, Iowa DNR, "Iowa, A Model for Energy Efficiency Programs," *Iowa Groundwater Assn.* 2 (December 1990): 12.

2   Patrick J. Deluhery, "Summary of Iowa's 1990 Energy Efficiency Act" (speech before NCSL's 1991 Annual Meeting, Orlando, August 1991).

3   Roya Stanley, Iowa Energy Bureau, October 1991: personal communication.

4   Iowa Department of Natural Resources (DNR), *Iowa at the Crossroads: 1990 Iowa Comprehensive Energy Plan* (January 1990): 4-1.

5   Paul Erickson, Iowa DNR, June 1991: personal communication.

6   *Iowa at the Crossroads*, 4–10.

7   Ibid., 4–11.

8   Iowa DNR, August 1991: personal communication with Kristin Rahenkamp, NCSL.

9   Ibid.

10   Ibid.

11   Ibid.

12   Teddi Barron and Tom Maze, *Iowa Motor Fuel Reduction Program: Final Report* (Des Moines, Iowa, June 30, 1989).

13   Ibid., 2.

14   Ibid.

15   Ibid., 106

16   Ibid., 147

17   Renew America, *Environmental Success Index* (Washington, D.C., 1991): 1.

18   Ibid.

19   Gordon Dunn, Iowa Public Utilities Board, October 1991: personal communication.

20   Joe J. Welsh, "Iowa's 1990 Energy Efficiency Act and Integrated Resource Planning," (speech delivered before the Pennsylvania General Assembly in Harrisburg, December 1991).

NEVADA

evada imports over 95 percent of its energy. An economic boom is responsible for increases in per-capita energy consumption over the last few years. Electricity use has increased substantially (**figure 34**). Nevada relies on petroleum and coal for its energy needs (**figure 35**). One northern Nevada utility experienced a 15 percent overall increase in customer electric use in 1991.[1] Gold mining continues to spur the growth, with a 52 percent increase in electricity consumption expected by 1993. Southern Nevada's Nevada Power, which leads the nation's electric utilities in customer growth, saw an increase of over 29,000 customers in 1989. Retired citizens and non-gaming industry add to the growth of Nevada.

State energy programs have been funded primarily through oil-overcharge settlements. As of June 30, 1990, Nevada had spent $12.9 million of the $16.6 million it received through these settlements.[2] As a condition of the settlements, the U.S. district court required that the funds be returned to consumers solely through the five traditional federal energy-assistance programs. These funds were invested through the Governor's Office of Community Services (the Nevada Energy Office [NEO]), the Department of Health and Human Resources, the Public Works Board, the Nevada Department of Transportation, and the Division of Aging Services, with most of the funds going to Low-Income Home Energy Assistance (LIHEA), the Nevada Department of Transportation, or the Weatherization Assistance Program (WAP).

While Nevada's approach to energy savings through IRP is considered one of the most innovative in the country, the state's approach to other energy

Figure 34

Total Annual Energy Consumption, Nevada, 1970–1988

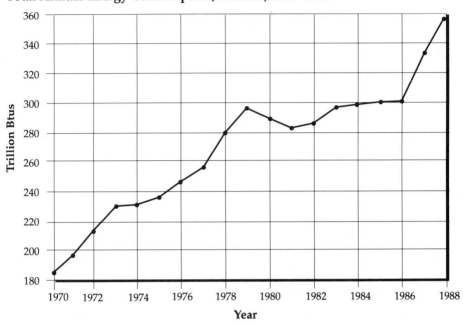

Source: Energy Information Administration, State Energy Data Report, Consumption Estimates, 1960–1988, DOE/EIA-0219, 1988.

programs is more traditional in nature, focusing on workshops and consumer education. One exception is the agricultural area.

The state has innovative agricultural energy programs in place. Particularly noteworthy is Nevada's Photovoltaic Agricultural Demonstration Project, which has funded remote solar-powered weather stations that are equipped with voice synthesizers to allow the device to "talk" with farmers over the telephone. The weather stations record climatic conditions; farmers then use the information to make optimum decisions about crop irrigation, crop selection, and fertilizer and pesticide applications. Each remote weather station records long-term weather data and is capable of executing sophisticated computer down-loading procedures. The Photovoltaic Agricultural Demonstration Project is also responsible to retrofit numerous abandoned windmill sites with cost-effective solar water pumps that allow water to be pumped to livestock and away from sensitive riparian habitat—an environmentally responsible option. While the photovoltaic technology is not new, the application is.

Figure 35

Nevada Energy Use by Source, 1988 (355 Total Trillion Btus)

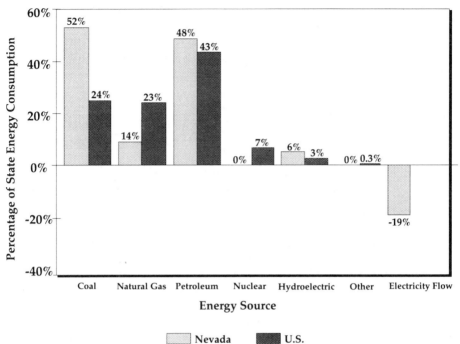

Source: Energy Information Administration, State Energy Data Report, Consumption Estimates, 1960–1988, DOE/EIA-0219, 1988.

The Governor's Office of Community Services is interested in 1992 in funding solar-powered electric fencing—an excellent tool for controlled grazing management. Time-control grazing is an effective way to develop more efficient and environmentally responsible herd management. Pastures are divided with movable, solar-powered fences so grazing can be staggered throughout the year, thereby reducing resource deterioration and loss of animal production. Studies have shown that the more grazeable units a manager has, the greater his flexibility to manage the resources.[3] Grazing management with solar-powered fencing also eliminates destructive over-grazing. Solar fences use a solar panel to charge a battery which, in turn, provides power to the fence-charging unit.

Nevada has considerable potential supplies of renewable energy, and legislative leaders have supported renewable energy tax credits. In 1991, the Legislature considered production tax credits for renewable energy technologies and the conversion to alternative fuels of vehicle fleets and other petroleum-fueled end-users. Many of these bills died in committee.

As a national geothermal leader, Nevada will continue to pursue geothermal resource energy production. The state also hopes to be a key research and production site for developing solar central-receiver technology. The Yucca Mountain, Nevada, potential nuclear waste repository site is one of Nevada's key energy-related issues. The site is being considered as the nation's first nuclear waste repository. Although no nuclear energy is produced within Nevada's borders, nuclear waste produced outside Nevada's borders is a major issue for Nevada legislators.

Energy Consumption by Sector

With the exception of Nevada's high energy consumption in the transportation sector, consumption in the traditional four sectors is close to the national averages. Nevada is known for its energy-intensive commercial sector in Las Vegas, yet commercial energy consumption is 18 percent of total state energy use, compared to the national average of 16 percent.

The Industrial Sector

Although it accounted for 24 percent (**figure 27,** page 135) of the state's energy consumption in 1988, Nevada's industrial sector is largely untouched by state energy programs. The Governor's Office of Community Affairs/NEO does manage a number of regional mining workshops. Energy-intensive gold and silver mining comprise a major share of the state's total industrial energy consumption. Twenty-five percent of the electrical load of one northern Nevada utility is strictly devoted to gold mining. Agricultural energy programs are managed by the State Conservation Commission, which has recently focused on photovoltaic applications for agricultural use. Because of Nevada's climate and geographic location, the state has the potential for future renewable-energy industrial applications.

The Transportation Sector

Transportation comprises a major share of total state energy consumption at 37 percent (**figure 28,** page 136). Nevada is geographically the seventh largest state, and NEO personnel claim tourists who drive to Nevada from California contribute to their poor ranking in this category. Reno and Las Vegas do not meet the EPA's clean air standards. The state is experimenting with compressed natural gas (CNG) as an alternative fuel to combat air quality problems.

The NEO runs car-care clinics, designates each October as car-care month, and uses these events to educate consumers about the environmental, safety, and energy-saving benefits of a well-tuned, well-maintained automobile. Reno and Las Vegas both have regional transportation systems. The Nevada Legislature passed legislation (AB 547) in 1991 that allows air-pollution control agencies to regulate traffic to improve air quality in the future. Ride-sharing and mass transportation use may be required of major Nevada employers in the next two years. Telecommuting workshops will take place in 1992.

The Residential Sector

The residential sector in Nevada consumed 21 percent of total state energy in 1988 (**figure 29,** page 137). The state has building standards that require energy-saving techniques; however, there is considerable room for improved energy savings in this area. The NEO conducts public information campaigns on energy efficiency and workshops on energy-saving building techniques. Many residential workshops managed by the NEO are sponsored by Nevada utility companies. Because of Nevada's recent population increases, numerous out-of-state builders do business in the state. New vacation home construction for Californians and others is expected to increase, and many Nevada residents believe there is an opportunity for energy savings in this sector. Nevada does prohibit electric resistance heating (NRS 523.360) in counties with populations above 100,000. Variances are required, and the NEO uses life-cycle costs to justify any variances. In 1991, the Nevada Legislature passed legislation (AB 359) that requires counties and cities to adopt building code changes requiring low-flow toilets in new homes and buildings by March 1, 1992. Ultra-low-flow toilets must be used as of March 1, 1993.

The Commercial Sector

Nevada's commercial sector in 1988 used 18 percent of the total energy consumed (**figure 30,** page 138). Casinos in the state use considerable electricity (the Mirage Hotel alone uses 25 MW to 30 MW of electricity annually); however, most casinos are considered efficient, since they have been the target of past energy-saving programs.[4] Because of the 24-hour nature of the casino business, air conditioning and heating loads on utilities are heavy and constant. When it comes to energy-saving programs, utilities are the major players in this sector. The NEO serves as a conduit between Nevada utilities and retail customers. The utilities have successful lighting rebate programs in place for commercial businesses. With total retail square footage in Las Vegas, including hotels, expected to increase by as much as one-third of existing square footage in the next two years, energy efficiency in this sector is important to NEO personnel.[5]

Energy-efficiency Statutes

The following list enumerates specific statutes and regulations related to incentives or requirements for energy efficiency.

NRS 334.070 Permits leasing of state facilities for energy conservation. This legislation allows the executive authority of any public body to enter into a contract to lease public property if the lease promotes the conservation of energy and promotes alternatives to traditional fossil fuels. The lessee agrees to construct a facility designed to conserve energy on the property and thereafter lease the property back to the governing body.

NRS 523.167 Prohibits electric resistance space heating. Counties with populations of more than 100,000 are required to prohibit electric resistance space heating. Variances are granted after life-cycle cost analysis by the Office of Community Affairs.

NRS 523.180 Requires that the design of any new building or addition to an existing building conform to the standards for the conservation of energy set forth in the chapter. Buildings covered include those used for residential purposes, educational, business, mercantile, institutional, storage, or public assembly. Buildings excluded are, among others, manufactured housing, greenhouses, and mobile homes.

NAC 319.800 et seq. Pertains to loans for conservation. These regulations allow low-interest loans to owners of single-family and multi-family homes, so long as the property is used as a primary residence. Nevada lending institutions participate in this process with state officials.

NRS 361.079 Exempts qualified systems for heating and cooling from taxation. "Qualified system" means any system, method, construction, installation, machinery, equipment, device, or appliance that is designed, constructed, or installed in a residential, commercial, or industrial building to heat or cool the building, or for water used in the building, or to provide electricity used in the building, by using solar or wind energy, geothermal resources, energy derived from the conversion of solid wastes, or water power. Any value added by a qualified system must be excluded from the assessed value of the building.

NAC 704.9491 Allows a utility company to recover costs incurred while implementing full programs for conservation and load management approved by the Nevada Public Service Commission. All prudent and reasonable costs incurred by a utility in implementing programs for conservation and load management that have been approved by the Nevada PSC are eligible for recovery through general rates. All costs must be readily identifiable and stringent record-keeping of all programs is required.

NAC 704.832 Requires a utility inspector to offer insulating jackets for water heaters and devices for limiting the flow of water through showerheads to eligible customers. Utility inspectors are required to offer to customers, at no charge, an insulating jacket for the water heater and a low-flow shower device, which limits the flow of water to a maximum three gallons per minute.

NAC 704.854 through NAC 704.858 Pertains to loans for purchase or installation of energy-conserving equipment. Among other things, these regulations allow Nevada utility companies to install energy-efficiency improvements for customers and permit repayment of the loan as part of the regular periodic billing for the utility's services. Utilities are required to charge reasonable and fair rates, and are not allowed to discontinue service if a customer fails to make loan payments on time.

Article, 10 Section 1, paragraph 8 of the Nevada Constitution. Allows tax exemptions for energy conservation. The Nevada Legislature may exempt by law property used for municipal, educational, literary, scientific, or other charitable purposes, or to encourage the conservation of energy or the substitution of other sources for fossil sources of energy.

Energy-efficiency Programs and Initiatives

Nonprofit Energy-efficiency Circuit Riders

In 1988, the NEO funded circuit rider positions for nonprofit agencies. Circuit riders are energy-efficiency experts—usually engineers—who travel across broad geographic areas to promote energy-saving measures with selected audiences. Nonprofit groups are similar in their energy consumption patterns to small commercial establishments and, since they are ignored by many state energy programs, represent a major area for potential energy savings. The circuit riders perform energy surveys, educate nonprofit management on the benefits of energy efficiency, and arrange technical engineering analysis and financing for nonprofit social service agency investments in energy efficiency. Unlike similar programs in other states that have been funded through oil-overcharge settlements, the circuit riders originally were funded through the Energy Extension Service program.

The NEO is responsible for the administration of the program. The circuit riders develop promotional materials and forms for measuring energy use patterns and savings potential for each nonprofit agency, and identify and contact as many nonprofit agencies as possible. The Office of Community Affairs supports this effort with seminars for nonprofit agencies to demonstrate the benefits of energy-efficiency investments.

When an agency requests assistance, the circuit rider visits the facility and makes recommendations for investments in energy efficiency. If a detailed engineering analysis is needed, the circuit rider arranges that through a participating utility or the Nevada Energy Management Institute (funded by the Office of Community Services). The agency can implement the recommended measures and arrange financing for the measures through below-market financing from area banks, or through the circuit riders' grant funds.

Utilities donate clock thermostats and insulation equipment to the nonprofit agencies and offer financial incentives such as rebates for energy-efficient lighting. As of August 1990, 97 of the 438 agencies contacted had asked for assistance from the circuit riders. Thirty-three audits were completed, with only minor analysis of actual energy savings. In the Reno area, for example, energy audit data for 14 agencies showed projected energy savings totaling close to $23,000. The total cost of energy retrofits for the 14 agencies was $48,000, which meant a simple payback of slightly more than two years.[6]

Agriculture Photovoltaic Demonstration Project

Nevada's Agricultural Demonstration Project was funded through oil-overcharge money in 1988. The project, responsible for the implementation of a number of energy-saving technologies, is managed by the Division of Conservation Districts, which reports to the State Conservation Commission under the Nevada Department of Conservation and Natural Resources. There are 29 conservation districts in Nevada, representing 3,000 farmers statewide.[7]

The project has concentrated on solar-powered agricultural applications. Three solar-powered remote weather stations have been funded through the project. These new high-tech weather stations are connected to phone lines with voice synthesizers that allow the remote station to "talk" or relay important climatic information to farmers, who call the stations for data that assists them to select crops, apply fertilizer or pesticides, and time crop irrigation. This new information allows farmers to optimize their agricultural efforts and ultimately reduces the amount of energy used in farming methods. The remote weather stations minimize economic risks for farmers and maximize more responsible decisions that affect energy use and environmental quality.

The rural nature and arid characteristics of Nevada present unique challenges for farmers and ranchers, especially with respect to delivery of stock pond watering systems. Through this demonstration project, ranchers were shown how photovoltaic stock pond watering systems were both economical and cost-effective as water delivery mechanisms. In the past, most stock pond watering systems were powered by windmills or diesel generators, both of which have high maintenance costs.

Photovoltaic systems used in this demonstration project saved energy and proved to have lower maintenance costs, as well as being less expensive than bringing new electrical lines into isolated areas. Old, abandoned windmills were retrofitted with new solar water pumps that produce eight gallons per minute, providing more than enough pressure to move livestock water supplies. The range of savings resulting from this program are diverse and include time, gasoline (for transportation), and costs associated with new electricity generation, transmission, and distribution. Prior to the development of the new solar application, water was trucked to livestock—an expensive, time-consuming, energy-inefficient, and labor-intensive process.[8]

The demonstration project combined environmental benefits with energy savings. New water systems were installed near cattle grazing areas to keep cattle away from environmentally sensitive riparian habitats, which were rehabilitated by the new technology. Because of the success of this project, Nevada's Sierra Pacific Power and the Electric Power Resource Institute (EPRI) will co-fund similar applications in 1992.

Community Energy Management

Boulder City, Nevada (population 12,000), a Las Vegas bedroom community, was faced with the need for new electricity production to meet the influx of new residents. Rather than purchase supplemental electricity with power costs estimated to be 15 to 25 times the price for existing hydropower resources, the city chose to implement community-wide, demand-side programs to increase the efficiency of existing electricity resources. Supplemental electrical costs through 2019 were conservatively estimated at $320 million. With an average annual population growth of 4.3 percent and an annual electrical consumption of 3.7 percent, the city has a short-term goal to decrease load growth to 3 percent per year.[9] By first addressing demand-side options, NEO personnel say Boulder City officials gained planning time and minimized the ultimate electricity costs to their customers.

Boulder City energy planners met with representatives of WAPA and the Nevada Office of Community Affairs to plan for increased energy efficiency in their city. In addition to hiring an energy specialist—an energy circuit rider—the city established an active customer education program to explain the need for conservation and the possibility of increased rates for purchase of supplemental power if conservation measures were not implemented. The energy circuit rider was responsible for publishing an energy newsletter. To encourage citizens to participate in conservation measures, the city also started a commercial lighting replacement program to save peak energy through more efficient lighting, and a residential audit program offering a free water heater blanket and up to $30 of weatherstripping materials. To discover energy-inefficient buildings, the city borrowed heat detection and load monitoring equipment under WAPA's equipment loan program. To save energy, street lighting was converted to high-pressure sodium lightbulbs, air conditioners were replaced with swamp coolers, windows were shaded, and heat pumps were installed. Through WAPA's Peer Match program, Boulder City was able to tap the expertise of the utility manager from Osage, Iowa, where successful similar demand-side programs were implemented.

Recent energy-efficiency measures by Boulder City include the implementation of an inverted block rate and establishment of a sliding scale hook-up fee. The inverted block rate assigns higher costs to those users who directly cause the city to purchase more expensive supplemental power from outside sources. In the spring of 1990, Boulder City introduced a new schedule of connection fees keyed to the relative energy efficiency of building design. Superinsulation, efficient lighting, heating and cooling, and solar water heating, among other measures, are rewarded, while high-demand energy buildings are charged a premium for electrical hook-up.

Nevada Energy and Environmental Education Network

Energy officials in Nevada recognize the undeniable link between environmental quality and energy production. In an attempt to educate Nevada residents about the environmental ramifications of energy production, the NEO started the Nevada Energy and Environmental Education Network (NEE-NET) in 1990. NEE-NET evolved from a previously successful energy program, The Northern Nevada Energy Project.[10] It is a unique program because of its focus on youth leadership in the energy field and its promotion by a broad-based coalition of private industry, public utilities, state and local governments, environmental organizations, and individual citizens.

NEE-NET is funded through a number of sources, including the University of Nevada, the Governor's Office of Community Services, the Rural Alliance of Superintendents (school districts), utilities, energy-related corporations, and private monetary and in-kind contributions. Private support has been crucial to the program's success. Ormat Geothermal, a Nevada corporation that created the first closed-system heat exchange unit, contributed funding in 1991, local banks contributed funds, and the XEROX Foundation provided in-kind printing and copying contributions to NEE-NET. The program is managed by the University of Nevada at Reno's College of Education.

Youth education dominates the program and includes camps, meetings, and educational projects. Through the program, the first National Energy Education Conference for kindergartners through 12th graders was held in Reno, Nevada, on February 28, 1992. The innovative program, best known for its Camp FLEEE (Future Leaders in Environment and Energy Education), sponsors two different leadership training camps that focus on energy. More than 450 youths attended Camp FLEEE during the summer of 1991. Registration fees for the week-long camps is $125 per student, with many underwritten by utilities, scholarships, private businesses, or school districts. The NEE-NET staff also produces a quarterly newsletter.

A day camp is usually sponsored for six days during June. NEE-NET personnel lend their expertise in rural cities to teach kindergartners through sixth graders the value of energy and environmental decisions. An average of 75 children participate with high school youth serving as teachers. The children learn how to clean up an oil spill, build a windmill (to learn about wind power), debate issues such as nuclear energy and CAFE standards, perform skits and plays, and make energy conversions between sources. They also visit a geothermal site and use solar cookers to prepare a lunch.

A second, week-long camp for approximately 100 fifth through eighth graders is usually held during August in the pine country north of Reno near Lake Tahoe. In the past, congressional representatives from California, Arizona, and Nevada have come to talk to the campers. This camp emphasizes "hard" science rather than traditional camp crafts or outdoor activities. Campers learn how to read an electric meter, how much energy is used to cook scrambled eggs or pancakes, how much energy they use in a normal day, and many other valuable lessons. Program directors focus attention on the economic and environmental consequences of energy decisions.

Until December 1991, the program was managed by a director of energy education in Reno, with one full-time equivalent support staff. A second director of energy education, hired in December 1991, will be based at the University of Nevada at Las Vegas (UNLV) to represent and concentrate on energy programs in southern Nevada. Staffing requirements fluctuate in the summer due to Camp FLEEE, when four teachers, seven high school students, and two college students are added to the payroll.

Integrated Resource Planning

Nevada has a projected electricity demand increase of 4 percent per year through the 1990s, due in part to the population growth of the Las Vegas area. The state has decided that energy efficiency is one of the best ways to meet future energy demands, and is relying on IRP—which has been mandated through legislation in Nevada (Nevada Revised Statute 704.746)—for its energy future.[11] IRP in Nevada began in 1983 when Jon Wellinghoff, of the Consumer Advocate's Office, brought the concept to two state senators who introduced an IRP bill that passed, leading to design of the first IRP plan in 1984. The Nevada IRP legislation and regulations are included in appendix 6.

In 1986, supply-side and demand-side options were contested by major utilities, and the 135-day response time mandated in the first law was found to be insufficient for analysis by state PSC staff. In 1987, the IRP law was changed to allow the Nevada PSC 180 days to respond to the plans, and the biennial plans required by the state were changed to allow the utilities three years between plans.

Two utilities are required to file under the IRP law: Nevada Power Company (NPC), which serves southern Nevada, and Sierra Pacific Power Company (SPPCo), which serves the northern part of the state, sell most of the electricity in Nevada. Many utilities spend as much as 4 percent of annual gross operating revenues on DSM programs; neither of Nevada's utilities, however, spend more than 1 percent of their annual gross operating revenues on DSM programs. (Appendix 7 contains a list of selected utility DSM budgets as a percentage of revenue.) Nevada passes one of the true tests for IRP: ratemaking includes recovery of costs for investments in IRP.

The 12 cooperative utilities operating within Nevada are not required to participate in the IRP process. The WAPA embarked on an IRP marketing and educational program that may change the status of cooperative utility participation in the IRP process in the western United States. Despite Nevada's reputation as a national leader in IRP, regulators, legislators, and utility employees interviewed for this publication agreed that demand-side programs are far from balancing traditional supply-side emphases in Nevada.

State regulators report that both utilities initially opposed IRP. Early IRP plans were submitted incomplete, and regulators say the utilities used the IRP process as a tool to slow energy-efficiency discussions, rather than to advance comprehensive planning strategies. However, the plans now are successful collaborations. Utilities share their information "in a workshop atmosphere, rather than the hearing room atmosphere present when the IRP process began."[12]

Because of differences in climate between southern (Las Vegas) and northern (Reno) Nevada, electrical use in these areas varies considerably. Air conditioning is used throughout the summer in southern Nevada, where the average residential consumption of electricity is close to 13,600 kWh annually and the average monthly electricity cost is $55. Retail rates are more expensive in northern Nevada, where the average annual residential consumption of electricity is only 7,500 kWh, yet costs an average of $53 per month.[13]

The lower cost of electricity in southern Nevada is due in part to the inexpensive hydroelectric supply generated by Hoover Dam. A winter 1990 survey found NPS rates among the lowest in the nation at an average of 5.3 cents per kWh.[14]

Nevada incorporates environmental externalities into electricity costs. The PSC drafted rules for quantifying the negative costs of electrical production. Part one of this study contains a copy of the externality values used by Nevada utilities. Questions remain about the reliability of demand-side options versus supply-side options in Nevada; demand-side options (conservation, energy efficiency programs) are seen by some as less secure.[15]

Table 26
Nevada Matrix

NEVADA Energy Program	Cost per kWh (cents/kWh)	Annual Savings ($)	Implementation	Environmental Benefits	Start-Up Costs	Fiscal Fairness (+,-)	Probability of Success	Benefit to Cost Ratio (Benefit/Cost)	Limitations or Special Circumstances	Confidence
Non-Profit Energy Efficiency Circuit Riders	Not available	$159,000 E (1)	Easy (2)	Medium (3)	Low (4)	(-) (5)	High (6)	2:1 (7)	Engineer salaries (8)	Low (9)
Photovoltaic Agricultural Demonstration Program	Not available	$3.6 million E	Moderate (10)	Medium (11)	Low (12)	(-) (13)	High (14)	1:10 (15)	Limited audience (16)	Medium (17)
Community Energy Management Program	Not available	Not available	Difficult (18)	High (19)	Low (20)	(-) (21)	High (22)	10:1 (23)	Public acceptance (24)	High (25)
Nevada Energy Network	Not available	Not available	Easy (26)	Medium (27)	Low (28)	(+) (29)	High (30)	N/A (31)	Results difficult to quantify (32)	Medium (33)

Matrix Notes

These notes correspond to the numbers in the matrix (**table 26**).

1) Annual savings figure assumes all 93 nonprofit agencies that requested assistance from the circuit riders received the assistance and achieved the same average savings as the Reno sample of 14 agencies.

2) This program has only minor administrative difficulties, due to the relatively low number of circuit riders and the relatively minor financial arrangements required for investments in energy-efficiency measures.

3) Environmental benefits deserve a medium ranking. Energy savings and the resulting decrease in emissions are moderate when compared to what could be accomplished if the energy circuit riders addressed energy-efficiency gains in Nevada's industrial sector.

4) Other states can duplicate this program with two circuit riders for approximately $100,000. Note: The more complex the engineering analysis required for energy audits, the higher the administrative costs to run this type of program.

5) The program is available only for nonprofit agencies. Other sectors are not allowed to participate.

6) The success of circuit rider programs in other states is well documented.

7) Benefit-to-cost ratio was calculated by dividing annual savings by the annual administrative cost to run the program.

8) Circuit rider programs are only as good as the circuit riders hired. Because of salary disparity with the private sector, it is often difficult to lure an advanced engineer into the public sector. Many times, circuit rider programs degenerate into public information and speaking tours, when engineering calculations and analysis are often more necessary.

9) No formal evaluations or documents were available for this program.

10) The moderate ranking results from a combination of factors, including advanced and changing photovoltaic technologies, remote locations, and traditional ranch stock pond systems.

11) Environmental benefits in the form of riparian habitat rehabilitation are unique and positive, but limited in scope.

12) Start-up costs are under $100,000.

13) The audience for this program is limited.

14) The program is easy to duplicate.

15) Costs of the program currently outweigh the benefits.

16) The limited audience for this program is a major factor.

17) No formal evaluations for this program were available. Interviews with NEO personnel and limited written materials were used for this analysis.

18) Any community-wide, energy-planning program that addresses the industrial, commercial, and residential sectors requires extensive labor to be successful. The organization, implementation, and evaluation associated with a community-wide energy program requires significant time.

19) Environmental benefits, in the form of reduced emissions, are high and immediate.

20) Start-up costs to the state government can be less than $100,000 if utilities and county and local personnel participate in the program.

21) The primary beneficiaries of this program are Boulder City residents, although all Nevada citizens theoretically paid for the program through utility payments and taxes. If the program were offered to all Nevada communities, the fiscal rating would be a plus (+).

22) Community-wide energy programs have proven successful in Wisconsin, Iowa, and other states.

23) The benefits of this program outweigh the costs.

24) In larger communities, as well as in Boulder City, public participation depends on adequate public information campaigns and perceived benefits. Generating public support for this type of energy program in all areas—especially areas with low electricity costs and no energy shortage—can take more than a year.

25) This program was a national award winner; written materials were abundant for this analysis.

26) Implementation of seminars, workshops, and educational meetings are relatively easy to administer. State energy offices have experience in this area.

27) Environmental benefits can be immediate, depending on the individual who acts on the information gained. Research shows that benefits from energy workshops for home builders accrue over the long run and are not immediate. NCSL suspects general energy workshop benefits accrue similarly for this program. Since there is no formal evaluation showing exactly when and how many people used what they learned, any judgment here is speculative.

28) Start-up costs for state government are under $100,000.

29) This program cuts across all sectors, and is available to virtually anyone who is interested.

30) Probability of success is high if local and state government and private-sector participation is maximized. Nevada's program had broad-based support.

31) No cost or benefit figures were provided to NCSL.

32) Coordinating statewide events can provide logistical challenges, but most state energy offices already have experience with such a network. As with most energy educational programs, it is difficult to determine when and if energy-efficiency improvements are actually implemented. Surveys to determine the results of the program—Has anyone used the information learned at the educational workshop?— are often conducted after significant time has passed and are not always accurate.

33) Confidence was rated medium. Energy programs of this type undoubtedly have benefits; however, accurate measurement of these benefits is virtually nonexistent. Future state energy programs like this will be more valuable if energy savings are actually documented over time.

Future Direction/Legislative Priorities

Nevada has no regular legislative session in 1992. In 1993, the Legislature will consider a number of energy-related proposals. The Nevada Legislature will consider business tax incentives and subsidies for alternative vehicle fuels. Nevada legislators will review existing building codes and proposals to strengthen them, especially for new residential buildings. Nuclear waste disposal issues surrounding the potential Yucca Mountain repository site will dominate the agendas of many legislators. Cogeneration legislation similar to Washington's 1991 legislation will probably be considered, as well as waste-to-energy legislation. Oil-overcharge funding for energy programs will be considered in 1993, as well.

Chapter Notes

1   James P. Hawke, NEO, January 1992: personal communication.

2   Nevada Governor's Office of Community Affairs, *Energy For Nevada: Report to the Legislature*, January 1991, 11.

3   Nevada Governor's Office of Community Affairs, *Energy Update* 2, no. 2, Summer 1991, 1.

4   Curtis Framel, Office of Community Services Energy Bureau, November 1991: personal communication.

5   Ibid.

6   Marilyn A. Brown et al., *Energy Efficiency in Nonprofit Agencies: Creating Effective Program Models* (TM-11602) (Oak Ridge, Tenn.: Oak Ridge National Laboratory, August 1990), 18.

7   Chris Freeman, Nevada Division of Conservation Districts, December 1991: personal communication.

8   Ibid.

9   Nevada Governor's Office of Community Services, "Nomination for the Administrator's Conservation and Renewable Energy Award," September 1991, 1.

10   Pamela E. Calhoun, University of Nevada, Reno, Research and Educational Planning Center, December 1991: personal communication.

     Much of the NEE-NET information was provided in December 1991 by Pamela Calhoun, University of Nevada, Reno, Research and Educational Planning Center. Additional information was obtained from the document *NEE-NET, Nevada Environmental Energy Education Network Energy Education Program*, published in June 1991 by the Energy Education office.

11   Tom Henderson, Nevada Public Service Commission, November 1991: personal communication.

12   Ibid.

13   *Energy for Nevada: 1991 Report to the Legislature*, 26.

14   Ibid.

15   Frank McCrae, Nevada PSC, July 1991: personal communication.

NEW YORK

ew York has been called the "most energy-efficient state" in the United States[1] partly because it uses the least amount of energy per capita, and partly because a large percentage of the state's population lives in the New York City area, where mass transit and multi-family dwellings are the norm. Additionally, over the past 15 years, the Legislature passed legislation directed at reducing the state's ultimate energy consumption, much of which resulted in significant energy savings, especially between 1973 and 1984 (**figure 36**). New York energy policymakers have also made significant investments in technologies to protect the environment; and many of these investments have improved energy efficiency.

New York's efforts to improve energy-use efficiency is attributable to the success of many initiatives operating in tandem to reduce the state's use of energy to drive economic activity. An important consideration that affects New York energy use is the composition of industry and demographic trends, which differs from state to state.

New York's approach to improving efficiency focuses on an energy-use model that considers all sectors and end-uses. Through a combination of government-sponsored energy-efficiency initiatives and utility DSM programs, all economic sectors have an opportunity to participate in energy-saving programs.

New York has a three-pronged approach to implementing energy-saving programs: the state Energy Research and Development Authority

Figure 36

Total Energy Consumption, New York, 1970-1988

Source: Energy Information Administration, State Energy Data Report, Consumption Estimates, 1960–1988, DOE/EIA-0219, 1988.

(NYSERDA), The New York State Energy Office (NYSEO), and the state PSC, all of which are influenced directly by the New York Legislature.

The state Energy Research and Development Authority is described in detail in section four. The NYSEO was formed in 1976 and is currently staffed by 250 employees. The NYSEO and the programs it runs save close to $1.8 billion annually in avoided energy costs.[2] The NYSEO develops state energy-policy initiatives; conducts state energy planning, in cooperation with the department of Environmental Conservation and the PSC; and participates in numerous policy studies, advisory bodies, and federal and state proceedings related to ongoing energy-management activities. The NYSEO also administers conservation programs that provide technical assistance and financial incentives to the residential, commercial, industrial, institutional, and transportation sectors.

The New York Public Service Commission (NYPSC) requires energy-efficiency improvements rather than construction of new power plants to meet projected increases in demand for electricity production. Guided by the State Energy Plan, the NYPSC set the ambitious goal to meet at least half its future energy needs through the year 2000 by using improved energy-efficiency techniques.[3] The NYPSC also has incorporated environmental costs into final energy costs, while leading the nation in IRP. Working closely with the NYPSC in many cases, the NYSEO formulated, implemented, and evaluated dozens of successful energy programs that have saved the state billions of dollars in avoided energy costs.

Why has New York pursued energy-efficiency gains? One explanation is that the total energy bill for the state is more than $28 billion per year, and the Assembly Energy Committee estimates nearly 50 percent of this amount is paid to out-of-state sources.[4] Some economists say the figure actually is larger, and drains the state's economy of approximately $1,000 per capita annually.[5] With the exception of hydropower, the state has few energy sources.[6]

Oil provides the majority of New York's energy supply. Despite the fact that New York is the fourth largest energy consumer of all states, it meets only 8 percent of its own energy demand and is dependent on foreign sources for 70 percent of its oil supply. New York utilities use more oil than utilities in other states.[7] In 1988, petroleum accounted for more than 50 percent of New York's energy supply (**figure 37**); as a result, the state's economy is more vulnerable to international petroleum supply interruptions.

After more than a decade of decreased energy consumption, accompanied by increased economic productivity, the latter half of the 1980s brought a disturbing trend to New York policymakers—increased energy-consumption patterns. For example, while the overall net energy consumption for the residential and commercial sectors has decreased since 1970, it increased 4.1 percent and 5 percent, respectively, each year since 1985, and by 1988 had returned to 1981 levels.[8]

New York was without an energy policy after the expiration in January 1984 of the State Energy Master Plan and Long-Range Electric and Gas Report. The Senate and Assembly were unable to agree on terms for new legislation to restore the planning process. In the absence of comprehensive energy legislation, Governor Mario Cuomo issued Executive Order (EO) No. 118, "Establishing An Energy Planning Process," on December 28, 1988, to establish an integrated energy resource plan for the State of New York.

The first state energy plan under executive order was released in October 1989. A draft update was released in July 1991, with the final plan expected in late 1991. The state energy-plan update reflects the basic premise that a sustainable energy future must support both a quality environment and a competitive economy. The plan, which has integrated energy policy planning with environmental and economic considerations, is detailed in section four.

Figure 37

New York Energy Use by Source, 1988 (3,586 Total Trillion Btus)

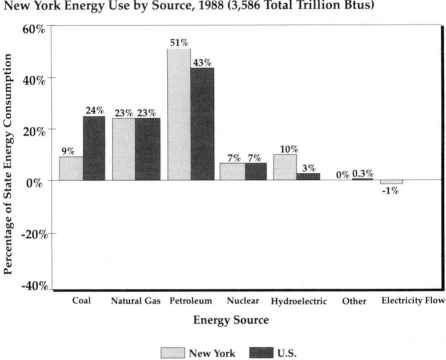

Source: Energy Information Administration, State Energy Data Report, Consumption Estimates, 1960–1988, DOE/EIA-0219, 1988.

Energy Consumption by Sector

New York's industrial sector has made the steadiest and most permanent progress toward reducing energy use in the state. The decrease in industrial energy consumption in the 1980s is explained in part by the loss of 300,000 manufacturing jobs between 1980 and 1988; and by the proliferation of service-related occupations, including insurance, finance, and real estate jobs.[9] The electric utility industry made tremendous strides to reduce its dependence on petroleum. The transportation, residential, and commercial sectors, on the other hand, have not matched the successes accomplished by the industrial sector.

The Industrial Sector

The industrial sector, which accounted for 19 percent of New York energy consumption in 1988 (**figure 27,** page 135), decreased oil consumption at an annual average rate of 8.7 percent since 1970. Public utilities also phased out the use of oil for electricity generation; the last new power plant fired primarily by oil was constructed in 1980.[10] The percentage of oil used by electric utilities decreased from 47.5 percent in 1973 to 28.9 percent in 1989.[11] The New York electric utility industry aggressively pursues energy-efficiency measures. The New York Energy Advisory Service to Industry program, which uses retired engineers to provide energy audits to industrial customers, successfully reduced this sector's energy consumption to save an estimated $64 million per year.[12]

The Transportation Sector

New York's transportation sector accounted for 24 percent of energy consumption in 1988 (**figure 28,** page 136). Densely populated urban areas and established mass transit systems explain the state's relatively low percentage of transportation energy use. Actually, transportation consumption of energy is nearing the 1973 peak level, and this sector made the least progress in overall energy-use reduction over the last 20 years. Since 1984, use of motor fuel has increased at an annual rate of 1.8 percent. Increased automobile fuel efficiency presents an opportunity for New York to improve energy consumption patterns in this sector.

Of the roughly 25 percent of primary energy consumption in New York used by the transportation sector, gasoline represents roughly 70 percent. In August 1990, Governor Cuomo announced a six-year, $40-million demonstration program to operate 250 cars, buses, and trucks with alternative fuels like methanol, compressed natural gas, and other alternative fuels. (The governor drives a methanol-powered car while in Albany.) The NYSEO also manages the Fleet Energy-Efficient Transportation (FLEET) Program, designed to assist fleet operators to implement fuel-saving measures, and the NYSERDA operates an alternative fuels demonstration program. The Legislature increased the state automobile fleet standards to 29 miles per gallon—1.5 mpg above the federal standard—by 1991 for the state's 6,500 automobiles.

The Residential Sector

Perhaps the most alarming trend to New York policymakers is the 22.8 percent increase in residential oil consumption between 1983 and 1988, in spite of the state's focus on increasing energy efficiency in this sector. Of the six states chosen for this report, New York's residential energy use (29 percent) is the highest (**figure 29,** page 137). This is explained in part by climatic conditions, but is due mostly to the number of energy consumers in the state.

New York policymakers rely on efficiency standards for new construction and utility-sponsored, energy-efficiency programs to curb residential consumption. An American Council for an Energy Efficient Economy (ACEEE) study estimated that electricity consumption for New York's residential sector could be reduced by 44 percent by using existing technology.[13]

The Commercial Sector

The commercial sector, accounting for 28 percent of New York energy consumption in 1988 (**figure 30,** page 138), has the highest percentage of commercial energy consumption in all states selected for this study. To address the problem of wasted commercial energy, the NYSEO in 1986 implemented the Small Business Energy Efficiency Program (SBEEP), a small but effective program that provides free energy audits to small businesses and recommends alternative energy products and efficiency improvements. SBEEP staff found that even if energy-efficiency investments were limited to those that would pay for themselves within two years, businesses already audited by SBEEP could save more than 16 percent of their annual energy costs. Lighting improvements and changes in HVAC systems may potentially halve the amount of energy used for a typical commercial business.

Energy-efficiency Statutes

Article 8 of the Public Authorities Law—The State Energy Office and the Energy Research Development Authority. In 1975, the Legislature created the NYSEO and NYSERDA. The mandate of the NYSEO is to coordinate energy policy for the long-term welfare of the state, to compile and distribute information regarding energy availability and prices, and to implement the various energy-conservation programs created by the Legislature. The mission of NYSERDA was to encourage the development of new energy technologies, with special emphasis on renewable energy sources and energy conservation. NYSERDA is described in section four.

Article II of the New York State Energy Law—The Construction Code. In 1977, the Legislature created an energy-efficiency code for new construction in New York. The code recognized the fact that lack of information, coupled to the desire to minimize initial costs in new buildings, usually prevented serious consideration of long-term energy costs when design decisions were made. Because it is more cost-effective to implement energy efficiency in the construction of a new building than it is to retrofit an old building, the code takes advantage of an important opportunity for energy savings. The code is periodically updated by the NYSEO; recent updates (the 1991 update is described in section four) focus on commercial buildings and are estimated to reduce consumption in affected buildings by 14 percent. The update includes the unprecedented development of standards for lighting equipment, including lamps and luminaries, installed in new commercial buildings. The Energy Code requirements for lighting equipment are the first to become effective anywhere in the nation and, like the New York ballast standards that became the model for national standards, could influence manufacturers, other states, and the nation.

Article 8 of the New York State Energy Law—Lighting Efficiency Standards. In 1979, to minimize the consumption of energy and provide for the efficient use of energy expended for lighting, the Legislature created lighting efficiency standards related to existing buildings. This was considered "cutting edge" legislation, since energy-efficient lighting technologies were not widely available in 1979. The legislation was directed at owners and lessors of buildings.

Article 17 of the New York State Energy Law—The Truth in Heating Law. Legislation passed in 1980 requires the seller of any real estate involving buildings to furnish the heating and cooling bills from the preceding two years to the potential purchaser before a contract is signed.

Chapter 858, L. 1977—The Home Insulation and Energy Conservation Act. In 1977, the New York Legislature passed the Home Insulation and Energy Conservation Act (HIECA), the first of its kind in the nation. This model legislation requires utility companies to offer free energy audits to homeowners and renters and to provide financing for the improvements recommended by the audit. Since its enactment, the program has been the cornerstone of residential energy-conservation services offered to New York residents. Nearly one million audits of homes or apartments and 3,000 audits of apartment buildings have been performed; the resulting improvements saved the equivalent of approximately 700 million gallons of oil. Since the program's inception, each dollar spent on the program has resulted in five dollars of energy savings, for a benefit-to-cost ratio of 5 to 1.

Section 5-108 of the New York State Energy Law—State Fleet Standards. The Legislature increased the efficiency requirement for new additions to the state car fleet to 29 mpg beginning in 1991. Previous requirements, first effective in 1981, required only 27.5 mpg. With 6,500 cars in the state fleet, approximately one-tenth of which are replaced each year, this requirement will save a minimum of 650,000 gallons of fuel during the period from 1991 to 1997; older, less efficient vehicles are cycled out of the fleet. After 1997, at least 150,000 gallons of fuel will be saved each year through continued use of higher mileage cars.

Section 66-c of the Public Service Law—Alternate Energy. In 1980, the Legislature enacted section 66-c of the Public Service Law, which established as state policy the conservation of energy resources and provided for their efficient use. The law encourages the construction of small-hydro, alternate-energy, and cogeneration power facilities by exempting them from some regulatory requirements and providing price stability for the electricity they produce. Under this law, the independent power industry added cogeneration and price competition to the state's electric industry. The Legislature also created a real property tax credit for residential solar and wind energy systems (Sections 487 and 487-a of the New York State Real Property Tax Law.)

Article 16 of the New York State Energy Law—Appliance Standards. New York has been a leader in appliance energy-efficiency standards. State laws regarding the efficiency of lighting and most household appliances have been in effect for many years. Compliance is monitored by the NYSEO and the attorney general's office. The unavailability of data from manufacturers makes accurate measurement of the energy saved by these standards difficult. However, the NYSEO estimates the cumulative energy savings attributable to these standards to be over $3 billion. The federal government has essentially pre-empted the New York standards, through the National Appliance Energy Conservation Act (PL 100-12), for room air conditioners and refrigerators manufactured on or after January 1, 1990. Refrigerators built after 1993 will be designed to consume no more than 700 kWh annually. By contrast, high-efficiency refrigerators commonly available today consume approximately 850 kWh annually.[14]

Chapter 659, L. 1989—Petroleum-Overcharge Funds. During the 1989 session, the Legislature passed the Petroleum Restitution Act of 1989, which distributed more funds to numerous energy programs that encourage energy conservation in all sectors. Similar legislation was passed in 1986 and 1987. Since 1986, New York has appropriated over $350 million of these funds.

Programs funded with these appropriations provide free audits and grants and loans for energy improvements to virtually all energy users in New York—homeowners, renters, farmers, businesses, industrial customers, schools, hospitals, local governments, and not-for-profit corporations. Although demand exceeded funding for most of the programs, the NYSEO estimates annual energy savings from them to be $1.16 billion.

The funds have also created two important research institutes. The Institute of Superconductivity at the University of Buffalo researches the commercial applications of high-temperature superconductors. The institute attracted over $8 million in research funds to the state. The Combustion Institute at Cornell University has become a leader in providing technical assistance to local governments around the state to deal with solid waste problems. Other research projects have been funded for various technologies that hold great potential for reducing long-term energy needs.

Energy-efficiency Programs And Initiatives

The New York State Energy Plan

On December 28, 1988, Governor Cuomo issued EO 118 to establish a formal state energy-planning process. The NYSEP (the plan) elevates three state agencies—the NYSEO, the Department of Public Service, and the Department of Environmental Conservation—to the status of "lead" agencies with the responsibility to devise and implement a state energy plan. In addition to implementing the plan, the lead agencies must update it at least once every two years to meet the state's changing energy needs.

Pursuant to EO 118, the plan, as well as its updated versions, will include a policy report on important energy issues, forecasts of every demand and supply assessment, and a supplement to the initial environmental-impact statement required by state law. Potential environmental impacts such as air emissions, solid waste, and water quality must be addressed. In addition, the plan requires agency staff to analyze present and future projections of fuel use, cost of energy services, economic development, and energy efficiency as part of a more comprehensive approach to energy planning.

In October 1989, the lead agencies, in accordance with the governor's EO, issued the New York State Energy Plan for 1989. The plan assessed current and future energy needs as well as their likely costs and specifically devised energy-planning strategies to meet New York's energy needs, taking into account the financial and environmental costs. The energy strategies touched on four general areas: energy efficiency, environmental quality, petroleum products and renewable resources, and electricity. A summary of the energy strategies incorporated in the plan for 1989 follows.

Energy Efficiency. The plan calls for improving energy efficiency by increased federal CAFE standards for automobiles and trucks, an improved transportation system to reduce vehicle miles traveled, increased state building efficiency codes and appliance efficiency standards, establishment of a target 20-percent reduction in energy consumption for state buildings by 2000, and pursuit of aggressive utility DSM programs.

Environmental Quality. The primary goal is reduction of air emissions of sulfur dioxide (SO_2), nitrogen oxide (NOX), and carbon dioxide (CO_2) from the combustion of fossil fuels. The plan's guidelines include adopting the California program of vehicle emissions standards; developing stricter emissions standards for SO_2 and NOX for existing and new power plants; and maximizing use of hydropower imported from Canada, where cost-effective and reliable.

Petroleum Products and Renewable Resources. Diminished dependence on imported oil and improved ambient air quality are important goals to New York. Cleaner burning domestic fuels would help New York achieve both these goals and, accordingly, the plan calls for the increased use of natural gas, especially as an alternative to electricity, and aggressive pursuit of new supplies of natural gas and other alternatives to oil.

Electricity. The plan emphasizes repeal of current state law that discourages procurement of new electricity generation, increased state jurisdiction over intrastate electricity transmission and wheeling, and strengthened current planning and siting processes with legislation to establish a process that focuses on the site-specific environmental impacts of a proposed electric-generating facility and reflects the findings and recommendations of the energy-planning and competitive-bidding programs.

It is not clear how the plan is enforced. The EO directs lead agencies to give deference to the plan with respect to their individual powers to make policy and implement programs. In theory, the plan is binding on all state agencies unless excused by state statute.

In July 1991, the new 1991 draft State Energy Plan was released that reaffirms and modifies, as appropriate, the 1989 plan goals mentioned previously.[15] The 1991 plan requires state and federal actions to reduce emissions from combustion of fossil fuels in buildings, transportation, and electricity generation. The 1991 plan recommends that the state:

1) Consider use of, and account for, environmental externality costs, to the extent possible in all aspects of state energy planning;

2) Integrate transportation and energy and environmental decisionmaking;

3) Integrate actions consistent with the plan's objectives to help reduce emission of greenhouse gases;

4) Support the need for actions at the federal level to ensure that energy prices reflect the real environment and security costs of energy;

5) Emphasize transportation programs that improve efficiency of existing systems and promote the use of more efficient modes; and

6) Place renewed emphasis on the promotion and demonstration of renewable resources.

The plan builds on the recommendations from the earlier 1989 plan. A brief summary follows.

Energy Efficiency. The plan reaffirms DSM energy-reduction targets of between 8 percent and 10 percent by 2000, and the 15 percent called for in the 1989 plan. To improve energy efficiency in the use of all fuels and lower total energy costs, New York is working toward a 2.5 percent per year annual reduction in the amount of primary energy consumption per dollar of gross state product (GSP).

Environmental Quality. The primary goal is cost-effective reduction of sulfur dioxide, nitrogen oxides, and carbon dioxide air emissions associated with energy consumption. Greenhouse gas emissions are to be stabilized at the year 2000 production levels, and further reductions of 20 percent are possible by 2008.

Renewable Resources. The draft 1991 plan calls for utilities to procure 300 megawatts (MW) of electric capacity from renewable resource technologies by January 1, 1994 (to be in service by 1998), in an effort to market test the cost-effectiveness of renewable resources.

Fossil Fuel Use. New York adopted the California low-emission vehicle program—as did 12 other northeastern states—to improve air quality and reduce petroleum use in the transportation sector. The plan calls for a 50-percent increase in the cost-effective and efficient use of natural gas for all sectors by 2008 and an increased demonstration and use of cleaner coal technologies for electricity production.

The New York State Energy Research Development Authority[16]

The NYSERDA is a public benefit corporation established by state legislation in 1975 and governed by a 13-member board of directors appointed by the governor with the advice and consent of the senate. The board is comprised of a not-for-profit consumer group representative, a not-for-profit environmental group representative, a research scientist, a gas utility representative, an electric utility representative, the chairmen of the New York Power Authority (the state's largest electricity broker) and the PSC, the state energy commissioner, the commissioner of environmental conservation, three public members, and an economist.

The energy authority is funded in part by annual appropriations from the state budget and employs a full-time staff of 90, 45 of whom are in research, development, and demonstration (RD&D). Funding sources include a fixed assessment on electric and gas utilities and the New York Power Authority. The energy authority also derives income from the investment of retained earnings and leased property, as well as from fees for issuing tax-exempt bonds and commercial paper. Its mission is to obtain and maintain adequate energy supply, accelerate the development of new energy technologies and foster research and development, especially in the energy-efficiency area.

The energy authority runs the Economic Development through Greater Energy Efficiency (EDGE) program to help industries achieve greater energy efficiency in their process technologies. EDGE offers financial assistance for demonstration of innovative energy-efficiency projects and targets industries. This successful program combines economic-development goals and energy-efficiency goals, while making New York industry more competitive.

Another innovative project run by the energy authority is the Energy Products Center, which encourages growth in innovative, energy-related products and services in New York. The center sponsors a lighting research center and provides technical assistance and financing to products that promote the effective and efficient use of energy and that provide jobs for New York residents. The state benefits from increased jobs and an expanded tax base if the energy enterprise succeeds. The center requires co-funding from the entrepreneur and/or other investors.

The 1990 Amendments to the New York State Energy Conservation Construction Code

The NYSEO has now completed a comprehensive update of its energy code, which became effective on March 1, 1991. Of particular importance in this update was the unprecedented development of standards for lighting equipment, including lamps and luminaries (lighting fixtures) installed in new commercial buildings. Until now, New York's energy code, like other energy-related codes and standards, has indirectly regulated lighting system efficiency. The code revisions turn attention away from power budget calculations, which proved less effective for ensuring efficiency and were difficult to enforce. Instead, the revisions include requirements for the minimum efficiency of all components of the lighting systems typically used for general lighting in commercial buildings; it covers fluorescent lamps, lamp ballasts, and luminaries. The requirements for luminaries were developed by a comprehensive review of over 10,000 luminaries offered for sale in the United States and included sensitivity analyses on specific design issues.[17]

The energy code requirements for lamps and luminaries will be the first to become effective anywhere in the nation and, like New York's ballast standards, could serve as models for manufacturers, other states, and national standards.

For commercial buildings that will have to add conservation features to meet code requirements, additional construction costs will average less than $1 per square foot but will pay back these costs in less than three years. Among all buildings for which changes are required, total energy consumption will be reduced 14 percent. On a statewide basis, this represents a savings to affected building owners of over $2.2 billion by 2000.

Energy-saving Programs

The energy-saving programs listed below were selected from hundreds of programs already implemented in New York, based on recommendations from New York state energy experts. The majority of programs are, or were, funded by oil-overcharge funds, utility rate surcharges, or the federal government.

The Energy Advisory Service to Industry (EASI) Program.[18] To assist small and medium-sized industrial-commercial companies identify and implement energy-conservation opportunities, the NYSEO in 1979 established the EASI program to provide free on-site energy surveys. The industrial sector consumes large quantities of energy, and the EASI program targets efficiency processes within this sector. The program focuses on energy efficiency; however, the environment has benefited from a reduction in fossil fuel use.

Specially trained EASI advisors, located in regional offices, work on site with industrial plant personnel to review utility, fuel and water-use data, and steam production needs. These advisors, often retired plant personnel or engineers with years of experience and training, can effectively communicate with the business owners. The advisor assesses a plant's opportunities to achieve savings in the energy-using systems, building structure and heating systems, and prepares a written report identifying no- and low-cost energy-conservation and efficiency improvements and cost-effective capital expenditures. Businesses that implement suggested improvements typically average 15 percent energy-cost savings.

Between 1979 and 1988, more than 8,300 EASI surveys were provided to New York businesses. As a result of the program, annual energy savings of roughly $64 million accrue to participating businesses. A study by an independent evaluation firm has shown a 20 to 1 benefit-to-cost ratio for the EASI program. The EASI program is part of a strategy to deal with the problems of acid rain, global warming, ambient ozone, and solid waste. EASI recommendations that reduce direct burning of fossil fuel potentially reduce annual emissions of CO_2 by 682,000 tons, SO_2 by 3,400 tons, and NOX by 1,400 tons.

The EASI program reaches a diverse and geographically-spread audience that ranges from a 15-employee lumber mill in rural New York to a 400-employee, high-tech computer plant in urban New York. While the future program potentially will be funded by investor-owned utilities and declining oil-overcharge funds, initial funding came from the federal government. The EASI program also receives in-kind support from selected contractors and local businesses.

The key components of the EASI program can be replicated in other states. All regional support centers use the same report format, and trained advisors meet the same quality control procedures and offer localized assistance. The number of regional support centers averages 10 statewide. For many owners of small and medium-sized businesses, the EASI program is their first form of education about the financial and environmental benefits of energy-efficiency improvements.

The Signal Timing Optimization Program (STOP).[19] STOP is designed to increase the knowledge of traffic engineers and technicians about the availability and use of state-of-the-art computer analysis for traffic signal timing and to increase the use of these programs. Traffic signal timing optimizes the flow of traffic through signaled intersections, while reducing idle time spent at stop lights and gasoline consumption by individual automobiles. This was the first statewide traffic-signal energy-management program of its kind in the Northeast.

The program offers training, technical assistance, and start-up financial assistance to local governments to help cover a portion of the initial cost of signal retiming. There are a number of proven computer programs available to improve traffic light synchronization. STOP demonstrates to local government personnel that advances in signal timing technology make changes cost-effective and easy to implement.

The target audience for this program is small—300 traffic engineering personnel from 77 local governments with populations of 10,000 to 100,000; five cities with populations of more than 100,000; and 11 New York State Department of Transportation (DOT) regions and one main DOT office.

This program is significant for two primary reasons. It provides broad program benefits to the general motoring public in the form of reduced emissions and direct fuel (gasoline) savings; and it addresses energy consumption in the transportation sector, within which it is often most difficult to reduce energy consumption. By training traffic engineers, STOP directly benefits the state's motorists.

The start-up costs for this program are not high. Annual budget cycles have averaged approximately $250,000, with the minimum award to a municipality set at $5,000 and the maximum award set at $82,500. The total award amount for any retiming project was established on a per-intersection basis of $800. The first budget cycle of approximately $300,000 saved more than $3.6 million in avoided fuel costs, and vehicle emissions were reduced as a result of the reduction in fuel consumption.

The Demonstration Furnace and Boiler Rebate Program.[20] This program was designed to determine whether a rebate would encourage the purchase of high-efficiency residential oil heating equipment. It was created by the state legislature in 1987 and funded with $1.5 million in oil-overcharge funds. The program's primary goal was to encourage consumers, either owning or residing in one- to four-family homes, to purchase oil-fired heating equipment with the highest Annual Fuel Use Efficiency (AFUE) rating available. The first round of the program ran from November 30, 1987, to December 15, 1988. It reopened in December 1989 and is still in operation.

To stimulate the purchase of heating equipment at the higher end of the efficiency range, rebates of $180 were made available to purchasers of oil-fired furnaces and boilers featuring an AFUE rating of 86 percent or higher (AFUE ratings range from 67 percent to 89 percent.). On average, $180 represents 35 percent of the cost differential between the higher cost, higher-efficiency heating systems and the lower cost, average-efficiency systems now sold.

Between November 1987 and December 1988, 6,425 consumers statewide qualified for a rebate, twice the number projected by the NYSEO.

The program was moderately cost-effective and generated approximately $2.09 in energy cost savings for every dollar invested. The benefit-to-cost ratio was more attractive on Long Island (2.77 to 1), where heating oil costs are higher than the statewide average. The average lifespan of the heating equipment was calculated at 35 years. Evaluators of this program made no adjustment for the declining efficiency of the equipment over its lifetime, and it was assumed that the equipment would be maintained.

The NYSEO estimates that, based on the efficiency of original equipment, consumers will save nearly $171 on their annual oil bill in addition to the $180 rebate. Of the $1.5 million spent on the project, $1.2 million went to consumers as rebates; the remainder went to administration, marketing, promotion, and advertising.

In terms of environmental benefits, approximately 245,000 pounds of SO_2, 138,000 pounds of NOX, and 173 million pounds of CO_2 were not dispersed into the atmosphere because of the purchase of high-efficiency heating units.

The Energy Investment Loan Program (EILP).[21] The Energy Investment Loan Program, managed by NYSEO, provides financial incentives to encourage businesses and multifamily building owners to invest in energy-efficiency improvements. The EILP provides interest subsidies on loans up to $500,000 made by private lenders for energy-conservation improvements. Any measure that results in energy savings from the alteration, repair, or improvement of a building or its equipment is eligible for the program. Subsidies on loans up to $1 million are available for on-site power production projects. The average loan request is $90,000. At the end of 1990, 78 New York banks, including Citibank and Chemical Bank, participated in the program.

The NYSEO contracts with lenders who provide loans at their market rate and an interest subsidy is issued to reduce the interest rate. The current interest rates are 4 percent for loan terms up to five years and 6.5 percent for loan terms up to 10 years. Each project must have a payback of not less than one year and not more than 10 years (except multifamily housing, which can have a payback of up to 15 years). Loan guarantees are available as an incentive for banks to loan to borrowers who do not meet normal credit underwriting criteria. Reduced energy costs that result from the energy project are factored into the credit analysis, allowing borrowers to meet strict financial criteria.

Through the end of 1990, over $57 million had been invested in energy-efficiency projects as a result of the $15 million awarded in interest subsidies. These funds were channeled to 562 New York state businesses, agribusinesses, and multifamily buildings. Annual savings of $8.7 million are expected to accrue through these investments. Since 1986, $33 million from New York's portion of oil-overcharge funds were appropriated to the EILP.

Future plans for the EILP include extending program benefits to alternative fuel vehicles, alternative fueling stations, and alternative energy sources. The governor's 1990–1991 Petroleum-Overcharge Plan included a proposal to modify the EILP from an interest subsidy program to a revolving loan program. The Legislature did not approve the bill before the session ended.

The Energy-Efficient Appliance Rebate Demonstration Program.[22] In 1987, the NYSEO introduced the Energy-Efficient Appliance Rebate Program to dealers and consumers in selected areas of the state. The program awarded a total of $5.6 million for more than 73,000 rebates ranging from $35 to $125 (50,000 rebates went to refrigerator purchases, and 23,000 went to air conditioner purchases). The program was funded through oil-overcharge funds appropriated by the Legislature and, like the furnace rebate program, was designed as a mechanism for expeditiously returning overcharge dollars to New York's energy customers, while encouraging the sale of energy-efficient appliances with the potential for economic benefits to both the consumer and the state through reduced energy costs and use. Another objective of the program was to demonstrate the effect of financial incentives on consumer appliance purchase decisions. Ninety-three percent of appliance dealers stated that the rebates influenced sales, with 65 percent indicating that it was a very influential factor.

The program ran from April 1, 1987, to May 21, 1988. The legislation that created the program limited participation to seven diverse counties, limited rebates to purchases of refrigerator-freezers and room air conditioners, and assigned the administration of the program to the state energy office. The rebate amounts ranged from 8 to 15 percent of the purchase price of the appliance. The program was marketed primarily at the point of purchase.

The program projects savings of $25 million in avoided energy costs during the lifetime of all appliances sold. This figure includes a reduction of 2.65 trillion Btus of energy that New York utilities did not have to produce. The program was cost-effective, with net energy cost savings of between $1.26 and $3.24 for every program dollar spent. The NYSEO estimated that the program reduced emissions of SO_2 by 1,069 tons, NOX by 362 tons, and CO_2 by 172,003 tons.

Note: Recent research suggests that appliance rebate programs may be less effective because of the less efficient appliances used in past state programs and the more stringent efficiency standards mandated by the National Appliance Energy Conservation Act (PL 100-12) for room air conditioners and refrigerators manufactured on or after January 1, 1990. Specifically, these appliances are becoming more efficient overall and appliances not meeting minimum standards are no longer manufactured. The trend toward higher efficiency appliances will continue, thus the cost-effectiveness of rebate programs is likely to decline over time. For example, refrigerators built after 1993 will be designed to consume no more than 700 kWh annually. By contrast, high-efficiency refrigerators commonly available today consume approximately 850 kWh annually.[23]

Transportation Systems Management Program (TSMP).[24] The energy office developed the TSMP to help commuters identify alternatives to single-occupant vehicle use. The program saves energy by reducing single-occupant vehicles, improving traffic flow, reducing peak period traffic congestion and air pollution, and encouraging the use of ride sharing and mass transit. The TSMP consists of a network of local transportation organizations that provide information and services to the public. Emphasis is placed on areas with specific transportation problems, such as a lack of downtown parking or areas congested by highway construction.

Program activities include establishing employer programs to assist commuting employees; encouraging formation of carpools; analyzing and promoting traffic mitigation measures; providing vanpool transportation services to employers, employees, and the elderly; and instituting labor recruitment programs. For example, major New York counties sponsor the *Commuters' Register*, a newspaper that uses easy-to-read maps and commuter ads for ride sharing. A person willing to share a ride simply checks the *Register* to find others who commute to and from the same locations. The cost of the ads is underwritten by the NYSEO and the New York DOT.

Areas served benefit from enhanced work opportunities, reduced employee commuter costs, decreased work force parking demand, and expanded labor markets for new employees. Reduced employee tardiness is also said to result from the program. New York analysts estimate the program has removed 9,300 single-occupant vehicles from the road, 717 tons of emissions from the air, and saves $3.5 million in fuel costs annually.

The Better Buildings Workshop Program.[25] In 1984, the NYSEO introduced the Better Buildings Workshop Program to acquaint remodelers, builders, and other housing professionals with innovations in energy-conserving building methods and products, and to encourage incorporation of these innovations into general building practices. The current program features state-of-the-art building techniques, practical super insulation, advances in heating and ventilation systems, and topical issues such as indoor air quality. Specific improvements later made by attendees included installing high-efficiency windows and attic insulation, insulating walls and foundations, wrapping buildings with an exterior air barrier product, and sealing electrical and plumbing penetrations.

Training programs are generally cost-effective due to their minimum expense and receptive audiences. The program's benefits actually accrue to people who never attend a workshop or read any of the books—the homeowners who realize the energy savings and other potential benefits (increased property values, more comfortable homes, more disposable income).

The benefit-to-cost ratio was estimated to be as high as 328 to 1— remarkably high. Workshops do reduce energy consumption, but the exact amount is difficult to estimate or measure. The NYSEO estimates that approximately 45,000 pounds of SO_2, 20,000 pounds of NOX and 14 million pounds of CO_2 were not dispersed into the atmosphere because of the use of energy-efficient procedures undertaken as a result of the workshop.

Integrated Resource Planning

Utility company energy efficiency in New York began in the early 1980s as a result of proceedings before the PSC about the Nine Mile Point 2 nuclear power facility. Interveners such as the Environmental Defense Fund argued before the PSC that investments in end-use energy efficiencies could offset the need for construction of the new facility. Consequently, the PSC evaluated the benefits of conservation programs (case 28223, May 7, 1982).

In December 1988, Governor Cuomo issued EO 118, instructing the energy office, the PSC and the Department of Environmental Conservation to establish an energy-planning process for statewide IRP for all fuels and all sectors. The resulting State Energy Plan called for an increase in energy-conservation programs. The plan was discussed in section four.

The PSC, in September 1987, directed utilities to prepare DSM plans that would include large-scale conservation and load management (C&LM) programs. Utilities are required to prepare biennial reports that are reviewed and approved by the PSC. New York utilities expressed concern that lost net revenue from end-use measures would decrease profitability and were therefore reluctant to implement C&LM programs unless new rate-making mechanisms were adopted to compensate for lost revenue.

In October 1991, the PSC directed each investor-owned utility to submit an IRP plan by May 15, 1992. The PSC subsequently granted interim adoption of incentive mechanisms that include rate incentive rewards, and the recovery of cost and lost revenues associated with C&LM programs. This followed the June 1990 order approving a revenue decoupling mechanism for a Rockland utility, similar to California's electricity revenue adjustment mechanism (ERAM) and the modified fuel revenue accounting method (MFRA) used by Central Maine Power.[26] It is anticipated that this new mechanism will be modified and designed to work effectively in New York.

Since the PSC initially directed utilities to begin development and implementation of conservation measures, utility spending on C&LM measures increased to $100 million in 1989, and to $150 million in 1990, resulting in a projected summer peak reduction of 356 MW. A total resource cost test, adopted by the PSC to measure the cost effectiveness of C&LM programs, includes an accounting of environmental externalities ($.0145/kWh).

Utilities are required to offer, at a minimum, these core C&LM programs:

1) Commercial and Industrial Lighting Efficiency Incentives Program

2) Commercial High-efficiency Space Conditioning Equipment Incentives Program

3) Commercial and Industrial Energy Audit Program

4) Demand Management Cooperative Program

5) Consumer Energy Information Program

6) Innovative Rate Design Program

Municipal utilities are not required to file an IRP plan in New York unless they generate their own power.

Utilities and state agencies jointly conducted a weatherization project to improve energy efficiency of low-income residences. A pilot program conducted by utilities and state agencies provided in-home assistance to elderly clients to determine their energy and housing needs and made available energy audits, weatherization, and home repair. Another pilot program, Power Partnership, offered low-income, payment-troubled customers home weatherization, energy education, and budget counseling. These programs were found to be ineffective and were not included in the core program requirements.

In addition to utility programs, the NYSERDA supports a variety of research projects on energy conservation, renewable energy, indigenous energy sources, energy recovery from municipal wastes, and environmental impacts.

Of the projected 4,028 MW of additional capacity needed by 2000, New York expects to produce 59 percent—2,370 MW—through improved efficiency. The PSC, through the State Energy Plan, has an IRP policy that relies heavily on four major initiatives—competitive bidding, "social externalities," DSM, and new profit incentives to encourage energy efficiency.[27]

Table 27
New York Matrix

NEW YORK Energy Program	Cost per kWh (cents/kWh)	Annual Savings ($)	Implementation	Environmental Benefits	Start-Up Costs	Fiscal Fairness (+,-)	Probability of Success	Benefit to Cost Ratio (Benefit/Cost)	Limitations or Special Circumstances	Confidence
Energy Advisory Service to Industry Program (EASI)	Not available	$64 million E	Moderate (1)	High (2)	Medium	(-) (3)	High (4)	40:1-E (5)	Conflicting annual savings totals (6)	Medium (7)
Signal Timing Optimization Program (STOP)	Not available	$3.6 million E	Easy (8)	High (9)	Medium	(-) 10	High (11)	21:1-E	Computer Technology (12)	Medium (13)
The Demonstration Furnace and Boiler Program	Not available	$441,000 E	Moderate (14)	Low (15)	High (16)	(-) 17	Low (18)	2:1-E (19)	A) Expensive B) Water heating impact ignored (20)	High (21)
The Energy Investment Loan Program (EILP)	Not available	$7.9 million E	Easy (22)	High (23)	High (24)	(-) 25	High (26)	2.4:1 (27)	Long payback periods (28)	Medium (29)

The commission requires open competition to develop the best option to provide additional generation or efficient alternatives to generation. Social externalities are included in all bidding plans. Factors such as emissions reduction, creation of jobs, and reliability are not usually included in costs but must be considered. In 1992, New York utilities are projected to spend nearly $150 million on DSM programs. Energy savings of 8 to 10 percent are expected by the year 2000 from these programs alone.

Matrix Notes

These notes correspond to the numbers in the matrix (**table 27**).

1) This ranking is primarily the result of the quality-control issues about implementation of the program in diverse, regional centers across the state. To set up regional offices and offer tailored, localized assistance like New York's requires significant labor and oversight. The implementation of this program would be easier if a state were to concentrate on a few major metropolitan areas, rather than on the entire state.

2) Environmental benefits are high because they are measurable and immediate. The goal of the EASI program is to increase the energy efficiency of companies that already use large quantities of energy. Emission reductions that result from this program are significant.

3) The monetary benefits from this program are concentrated in the industrial and commercial sectors; the residential and transportation sectors are not addressed.

4) Massachusetts has replicated the EASI program structure with successful results. All EASI advisors use the same report format, which can easily be duplicated.

5) The only outside evaluation NCSL could find was done in 1987 by Pete, Marwick, and Main, who calculated the 40 to 1 ratio.

6) NCSL discovered discrepancies in the annual savings figures. Published figures show annual savings of $64 million, while NYSEO officials quote annual savings of $42 million. NYSEO officials noted that the techniques used to figure the $64 million in annual savings might be flawed. In any case, savings are substantial when compared to other programs.

Complex technical modifications recommended to industrial clients are sometimes not implemented, due to the fact that a more detailed engineering analysis is required than is provided by the EASI program.

7) Program materials and reports were available for this evaluation. NCSL did not receive a formal program evaluation of the EASI program.

8) The program is easier to implement than most state energy programs because of its narrow audience, the limited number of parties that participate in STOP, the tested computer programs, and the manageable workshops (workshops are characterized by interested, receptive audiences).

9) The direct fuel savings and associated emission reductions contribute to an extremely high rating for environmental benefits.

10) All state municipalities are eligible to participate in the program, and state motorists receive economic benefit directly through reduced fuel consumption and improved traffic flow.

11) California, Virginia, Massachusetts, and Iowa, among others, use traffic synchronization to achieve similar benefits. The technology is accepted and proven.

12) There are many computer-based traffic synchronization programs available (TRANSYT-7F and Passer II-84, for example). A program's success often depends on the willingness of municipal engineers to implement new technologies—some that are not traditional transportation engineering concepts. Because of a number of well-documented institutional and personal idiosyncrasies, new computer applications are sometimes accepted slowly.

13) Program materials and reports were available for this evaluation.

14) Because of the extensive prior communication necessary with manufacturers, distributors, and installers to acquaint them with the program, the implementation of this program might be more difficult than expected. The advertising and marketing components of this program require extensive labor.

Note: The NYSEO discovered the most effective way to use the heating equipment and interest consumers in the program was to direct their efforts at the heating equipment retailers.

15) Environmental benefits are low in this program, due in large part to a payback time of 16 years on the original investment. There is a reduction in air emissions, but it is not substantial or immediate.

16) Generally, rebate programs of this type are expensive in the short term. To make the program effective, large sums often must be appropriated so that the rebate is large enough to attract consumers, but not so large as to make the cost-savings negligible.

17) The program was available statewide; however, the number of rebates awarded varied in each utility district from 60 percent to 3 percent. This statistic leads NCSL to believe that the outreach efforts might not have been as uniform as desired. In addition, not all homes in New York use oil-heating equipment, so many homeowners were not included in the program.

18) New York has a disproportionate number of residential oil furnaces in the state; some estimate that New York has 50 percent of the U.S. total of these furnaces.[28] Other states could not expect New York's high participation rate.

19) Savings would increase substantially if the NYSEO evaluation had calculated savings by comparing the energy use of the equipment being replaced with that of the new equipment being installed. Using this scenario, the calculated savings would increase from $3 million to $15 million, even with the free-rider factor. As the NYSEO reports, this method would not have been appropriate, since most participants would have purchased a system more efficient than their old system. The free-rider factor takes into account those who benefitted from the program without paying or formally participating in the program. Free-riders would have taken the desired action even if the program were not available.

20) The NYSEO did not factor the water heating effect that resulted from this program, because no reliable formula existed to do so. Of the program participants, 67 percent indicated that their heating equipment is also used to heat water. The NYSEO is claiming only 42 percent of the savings, so their cost savings are conservative. The remaining savings are dismissed, since program participants said they either would have purchased the unit without the rebate, or the rebate did not influence their purchase decision (free-riders).

21) NCSL received formal program evaluations of the program.

This program was expensive to start; $1.5 million was required over two years to run the program.

22) The NYSEO contracts with lenders to provide loans to New York businesses. The NYSEO does not serve as the "bank," which makes the administration of this program easier than most. Since the program is run through the state's Energy Conservation Program, many other states would already have the administrative support in place to start this program.

23) Energy-efficiency projects implemented through this program have immediate environmental benefits in the form of reduced air emissions from avoided electricity production.

24) Start-up costs are high. With an average loan amount of $90,000, "buying down" (or subsidizing) only 100 average loans would cost $720,000 (100 X 90,000 X 8 percent) alone, without administrative costs. Since subsidized interest rates for energy loans are currently around 4 percent for loan terms up to five years, and 6.5 percent for loan terms up to 10 years, NCSL arrived at the 8 percent figure by subtracting a subsidized average interest rate (5 percent) from the market rate (13 percent). This additional amount would need to be available to fund only 100 projects in the first year.

25) This program is not available to all citizens, and the oil-overcharge money was appropriated to New York on the basis of total oil consumption by the general public. Therefore, those who paid earlier do not directly benefit from the program.

26) Programs like this are easily duplicated.

27) Calculated by simply dividing additional investments in energy efficiency (benefits) that resulted from original loan subsidies (costs).

28) Paybacks up to 10 years are allowed for businesses, and up to 15 years for multifamily housing. The interest subsidy is available for 10 years or the term of the loan, whichever is less. Many energy-efficiency investments in multifamily dwellings take more than 10 years to recover. To ensure optimum use of limited state funds, policymakers might consider shorter payback periods.

29) There was no formal program evaluation available for this program. Interviews and NYSEO's *1990 Report on Use of Petroleum Overcharge Restitution Funds* were used for much of this analysis.

Future Direction/Legislative Priorities[29]

The following legislative proposals probably will be considered by the New York State Assembly Energy Committee in the 1992 legislative session.

1) A sales tax on new automobiles that fall below the state average fuel economy standard, and a reduction in the sales tax on new automobiles that greatly exceed the standard.

2) A Home Energy Rating System (HERS) and a requirement that hook up fees charged by electric utilities be geared to the energy rating of a home.

3) A ban on the use of electric resistance as the primary heat source in new construction.

4) Minimum efficiency standards for power plants.

5) A requirement that any public customer of the New York Power Authority not participating in a DSM program by 1993 shall not qualify for renewal of power contracts.

6) A requirement that 20 percent of vehicles purchased for the state's fleet be powered by fuels other than gasoline by 1995, subject to availability.

7) An instruction for the State Energy Research and Development Authority to issue bonds to finance state facilities' energy-efficiency programs.

8) A provision that any public agency will be able to retain 50 percent of the savings achieved through participation in energy-efficiency programs.

Chapter Notes

1   Public Citizen, a press release announcing *Energy Audit: A State-by-State Profile of Energy Conservation and Alternatives*, October 1990.

2   NYSEO, *1989 Annual Report to the Legislature and Governor*, January 1990, 2.

3   Laurence B. DeWitt, "Energy Plan Brings New Electric Era in New York," *Forum for Applied Research and Public Policy* 5, 4 (Winter 1990): 33.

4   New York State Assembly Energy Committee, "New York's Persistent Energy Crisis: The Consequences of Overuse and the Availability of Power Through Energy Efficiency," (Albany: January 24, 1991), 8.

5   Dewitt, 32.

6   W. Conrad Holton, "Energy Policy in the Aftermath of Kuwait," *Empire State Report Magazine* (April 1991): 20.

7   Arnold R. Adler, Legislative Commission on Science and Technology, State of New York, August 1991: personal communication.

8   New York State Assembly Energy Committee, 6.

9   Arnold R. Adler.

10   New York State Assembly Energy Committee, 8.

11   Ibid.

12   NYSEO, *Description of Program: Energy Advisory Service to Industry (EASI)*, 1990.

13   Peter Miller et al., "The Potential for Electricity Conservation in New York State," American Council for an Energy Efficient Economy for the New York State Energy Research and Development Authority, 5–8.

14   Ruth Horton, NYSEO, October 29, 1991: personal communication.

15   NYSEO, *New York State Energy Plan Draft Update*, August 1, 1991.

16   This information was compiled from the NYSERDA *Annual Report 1989–90* and *Toward a Secure Energy Future: A Multiyear Research, Development and Demonstration Plan, 1990-1995*, NYSERDA, Fall 1990.

17   Ruth Horton.

18   Information obtained from informal program evaluation and *Description of Program: Energy Advisory Service to Industry (EASI)*, published by the NYSEO, 1990.

19   Ibid.

20   Some of this information was obtained from a formal program evaluation of the New York State Demonstration Furnace and Boiler Rebate Program by the Evaluation Unit of NYSEO, March 1990.

21   Some of this information was obtained from the *1990 Report on Use of Petroleum Overcharge Restitution Funds*, NYSEO.

22   Some of this information was obtained from a formal program evaluation of the New York State Energy Efficient Appliance Rebate Demonstration Program by the Evaluation Unit of the NYSEO, May 1989.

23   Ruth Horton.

24   NYSEO Fact Sheet, Transportation Systems Management Program.

25   Some of this information was obtained from a formal program evaluation of the Better Buildings Workshop Program by the Evaluation Unit of the NYSEO, April 1990.

26   James Gallagher, NYPSC, November 1991: personal communication.

27   Dewitt, 33-34.

28   Arnold M. Adler.

29   New York State Assembly Energy Committee; supplemented with personal communication with the late Bill Hoyt, chair of the State Assembly Energy Committee, August 1991.

WASHINGTON

ashington is the largest and fastest growing of the Northwest states. The state's population increased from 3.4 million in 1970 to 4.8 million in 1990, with total employment doubling from 1.1 million in 1970 to 2.2 million in 1990.[1] Annual per capita electricity consumption in Washington declined by more than 700 kWh between 1979 and 1989; and between 1972 and 1988 Washington businesses used 3.7 percent less energy per dollar of economic output.[2] Total energy consumption continues to increase, however, due to rapid economic expansion and an increase in population (**figure 38**). Unlike most states, Washington relies on Alaska for 85 percent of its oil. Despite the availability of relatively inexpensive electricity, Washington's total annual energy expenditures in 1990 surpassed $9 billion.[3]

Washington's economic growth has depended on low-cost energy. Some argue that inexpensive hydroelectric power from the Columbia River Basin and the fact that the largest of the five federal power marketing administrations, Bonneville Power Administration (BPA), supplies the Northwest with nearly 50 percent of its electricity make the state less likely to be interested in energy efficiency; electricity rates in the Northwest are the cheapest in the nation. Hydroelectricity supplies almost 40 percent of Washington's energy (**figure 39**), making the state unique—no other state provides as much electricity through this renewable resource.

Much of the state's industry developed because of plentiful supplies of inexpensive electricity. At the request of the federal government during World War II, major aluminum companies built smelters in the Northwest to take advantage of the region's low-cost hydroelectric power.

234

Figure 38
Total Annual Energy Consumption, Washington, 1970–1988

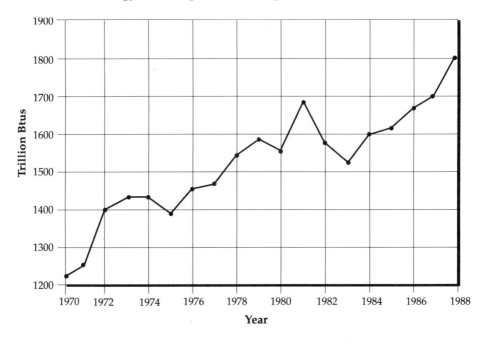

Source: Energy Information Administration, State Energy Data Report, Consumption Estimates, 1960–1988, DOE/EIA-0219, 1988.

Natural gas use in Washington is expected to increase in the 1990s, not only in its traditional uses of building and industrial process heat, but also in end-uses like electricity generation and transportation. The low cost, abundance, relative ease of delivery, and environmental advantages make it particularly attractive to Washington policymakers.[4] Washington produces no natural gas; however, the state benefits from its proximity to substantial natural gas reserves in western Canada. The growth in natural gas use was 12 percent per year between 1984 and 1989. No other major fuel in Washington approximates this growth rate.

The Pacific Northwest is known for its regional electrical power planning approach and its activities to reduce carbon dioxide emissions as part of a comprehensive global climate change initiative. Washington legislators are interested in how the quantification of environmental externalities (that result from electricity power production) will be treated by Oregon policymakers, who are evaluating this issue in 1992. The Oregon PUC opened a docket on the treatment of environmental externalities in 1991. Under a mandate, the Washington State Energy Office (WSEO) will deliver environmental externality suggestions to the governor in late 1992.[5]

Figure 39

Washington Energy Use by Source, 1988 (1,807 Total Trillion Btus)

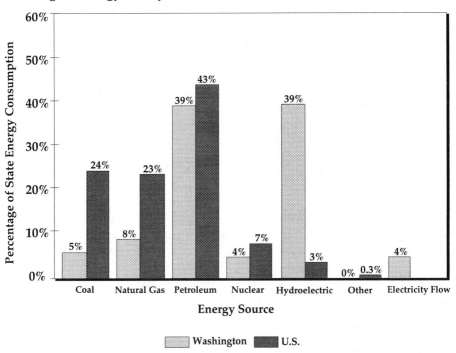

Source: Energy Information Administration, State Energy Data Report, Consumption Estimates, 1960–1988, DOE/EIA-0219, 1988.

Washington faces special risks from global climate change, since its economy depends on healthy forests, maritime industries, agriculture, and inexpensive hydroelectric power.[6] Rather than wait for more scientific evidence to accrue, Washington has decided that enough evidence exists to warrant stabilization of carbon dioxide emissions at 1988 levels by 2000. Although there is no formal legislation in place to limit carbon dioxide emissions, there *is* an informal policy. The new state energy strategy, due for release in early 1993, is likely to address carbon dioxide emissions. Petroleum is responsible for the majority of Washington's energy-related carbon dioxide emissions (58 percent), with coal responsible for only 1 percent, electricity generation for 23 percent, natural gas for 9 percent, wood fuel for 7 percent, and other sources for less than 1 percent. The transportation sector is responsible for 41 percent of the total carbon dioxide emissions in Washington.

The Washington Legislature encourages energy efficiency in new home construction. Results show that energy-efficiency investments are both cost-effective and less damaging to the environment than new sources of electricity generation.[7]

The Legislature also passed significant legislation in 1991—the most noteworthy of which was ESHB 5245—that establishes the framework for a state energy strategy. The legislation, signed into law by the governor, directs the WSEO to establish an advisory group and use a collaborative process to develop a state energy strategy during 1992, calls for increased conservation measures in state facilities and schools, and formally declares that cogenerated power from state-owned facilities is an energy-saving priority for the state.

The state energy strategy must consider all forms of energy and, to the extent possible, quantify the costs and benefits—including economic and environmental—of energy alternatives. With respect to energy savings in state facilities, the new legislation allows a state agency, once it identifies energy-saving measures, to "sell" its savings to Washington utilities, which in turn are allowed to credit energy bills or compensate the agency in other ways. Section four contains a detailed explanation of this legislation. Since Washington is seen as a leader in energy efficiency, similar legislation is expected to surface in other states in 1993.

Energy Consumption by Sector

The Industrial Sector

Washington's industrial sector consumes 35 percent* of the state's energy supply annually (**figure 27**, page 135). The industrial sector includes agriculture, metal refining and processing, paper manufacturing and pulp mills. BPA, which generates 60 percent of the state's electricity, until 1979 gave rates that allowed healthy profits to industrial accounts. Aluminum companies use a tremendous amount of electricity in Washington. Between 1979 and 1986, Northwest aluminum companies' rates increased by 900 percent. Energy costs, which were usually 3 percent to 5 percent of the costs of producing aluminum, grew to approximately 30 percent of total production costs.[8]

The primary metals industry, led by the aluminum companies, has increased energy use since 1978—due more to declining world aluminum prices than to decreasing energy efficiency. Oil remains the primary energy used by the industrial sector, followed by natural gas and electricity.

Prior to the OPEC oil embargo in the early 1970s, industrial energy use in Washington increased steadily; oil supplied most energy needs. With the rise of oil prices in the 1970s, industry dependence on electricity grew

*Note: The WSEO estimates energy consumption in this sector to be 25 percent.

while use of petroleum and natural gas declined.[9] At the same time, the value of aluminum fell on the world market. Despite losing nearly 10,000 jobs in Washington aluminum plants and watching aluminum revenues to BPA drop from 22 percent to 14 percent in 1986, the aluminum industry is still attempting to compete with Japanese, South American, and Middle Eastern companies whose energy costs are as little as one-third that of Northwest smelters.[10]

The Transportation Sector

The Washington DOT estimates that vehicle miles traveled (VMT) doubled between 1970 and 1990.[11] The transportation sector consumes 28 percent* of Washington's energy annually (**figure 28,** page 136). During the 1980s, VMT grew nearly four times faster than the 12.7 percent growth in the state's population and is expected to continue to grow an additional 25 percent by 2000.[12] The number of vehicles operating in the state is expected to grow from 4.1 million in 1990 to 5 million in 2000. Seventy-five percent of each barrel of oil consumed in Washington goes to the transportation sector; Washington's citizens spend $2 billion annually to fuel motor vehicles.[13]

Over the past 10 years, the average fuel efficiency for Washington vehicles has improved 29 percent, from 13.6 mpg to 17.6 mpg. New passenger vehicles achieve an average of 28 mpg. Savings associated with the increase in mpg saved the state $1 billion per year in energy costs and required 40 fewer oil super tanker trips into the sensitive Puget Sound area.[14]

State policymakers have successfully addressed transportation energy consumption through a number of strategies, including a successful telecommuting program and transportation demand management legislation that is discussed in section three.

The Residential Sector

This sector accounts for 20 percent** of the state's total annual energy use (**figure 29,** page 137). Strong state residential building codes have increased the heating efficiency of new homes by 60 percent since the late 1970s. Improved insulation levels in residential buildings were responsible for energy-efficiency gains in the 1980s, after the Legislature enacted a residential building energy code in 1978. The code was strengthened in 1980 and again in 1985, and is responsible for saving 67 average megawatts (aMW) of electricity in the 1980s, or enough electricity to power Spokane, Washington, for a year. In 1990, the Legislature again improved the building energy code to make it one of the strongest in the nation.

*Note: The WSEO estimates energy consumption in this sector at 50 percent.
**Note: WSEO estimates energy consumption in this sector at 15 percent.

Figure 40 shows that residential energy consumption rose steadily until 1979, when high energy prices and a failing economy contributed to a three-year decline in residential energy consumption. When the economy began to rebound, consumption of residential energy also rose, with electrical use in the lead. Despite the overall increase in residential energy consumption, individual households have reduced their energy use in response to higher prices, more stringent energy codes, and the availability of conservation programs.[15] Since 1972, average household energy use fell by over 20 percent. The 1986 Washington State Energy Code (WSEC) contributed to this trend. The average household in Washington used about 130 million Btus for heat, hot water, lights, and appliances in 1970. By 1990, that total had dropped to 98 million Btus. Improvements in appliance efficiency, reduction of average household size from 2.9 people in 1970 to 2.4 people in 1990, and improvements in space and water heating equipment contributed to this trend.[16]

Figure 40

Yearly Energy Use in Washington's Residential Sector

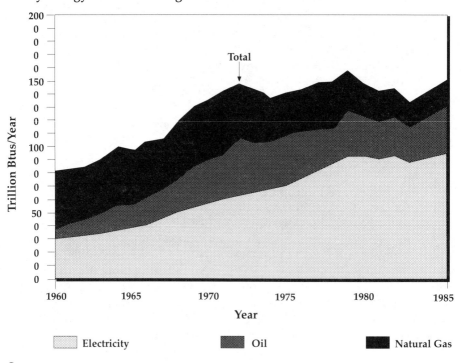

Source: Washington State Energy Office, December, 1986.

The Commercial Sector

Washington's commercial sector, consisting of office buildings, retail and wholesale stores, restaurants, institutional buildings and government facilities, consumed 16 percent* of the state's total energy in 1988 (**figure 30, page 138**). In 1960, the commercial sector accounted for less than two-thirds of the jobs in Washington; by 1985, its share had increased to more than 80 percent. Overall energy consumption in the commercial sector has grown steadily since 1970, with most of the increased energy demand met by electricity. Nearly 75 percent of the commercial sector's total energy expenditures are on electricity.[17]

Figure 41 shows the growth of energy consumption in this sector since 1960, use of electricity and natural gas grew between 1970 and 1990, while use of petroleum remained constant. Although energy use per employee declined for petroleum and natural gas, the ratio increased over the last 20 years for electricity because more businesses use office machinery— computers, copy machines, and other information processing equipment.[18]

Figure 41

Yearly Energy Use in Washington's Commercial Sector

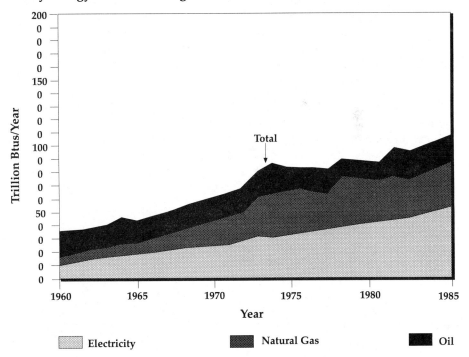

Source: Washington State Energy Office, December, 1986.

*Note: The WSEO estimates energy consumption in this sector at 10 percent.

Energy-efficiency Statutes

Key legislative data for Washington state energy-efficiency building measures is included in appendix 8.

Section 19.27A.010 - 19.27A.120. Establishes minimum thermal and energy standards for new residential and commercial buildings in Washington. Amended in 1990 to establish the most aggressive residential standards in the country and in 1991 to establish commercial standards in excess of current American Society of Heating, Refrigerating, and Air Conditioning Engineers (ASHRAE) standards. This legislation required all cities, towns, and counties to enforce the revised state energy code. The state building code council is required to follow the Legislature's guidelines set forth in this section to design a revised code that requires new buildings to meet a certain level of energy efficiency, but allows flexibility in building design and construction within that framework. The council is required to consider possible health and respiratory problems caused by insulating buildings so tightly that the rate of air-exchange is significantly retarded.

Section 28A.58.441 (2). Authorizes school districts to use capital project funds for energy audits and implementation of energy-conservation measures. Washington legislators defined an energy audit in 1981 as a survey of a building or complex to identify the type, size, energy use level, and major energy-using systems; to determine the appropriate energy-conservation maintenance or operating procedures; and to assess any need for the acquisition and installation of energy-conservation measures, including solar energy and renewable resource measures.

Section 35.92.105. Permits municipalities to issue revenue bonds or warrants for the purpose of funding energy conservation measures. In 1981, this legislation enabled a city or town to issue revenue bonds or warrants to defray the cost of financing programs for the conservation, or more efficient use, of energy. The bonds or warrants are deemed to be for capital purposes within the meaning of the uniform system of accounts for municipal corporations.

Section 39.35A.010. Permits the state and municipalities to enter into performance-based contracts with energy services companies in order to develop energy -efficiency projects with no initial capital outlay (1985). The Legislature found that performance-based energy contracts were a means by which municipalities could achieve energy conservation without capital outlay. Each municipality is required to publish in advance its requirements to procure energy equipment and services under a performance-based contract. The announcement shall state concisely the nature of equipment and services for which a performance-based contract is required, and shall encourage firms to submit proposals to meet these requirements.

Section 39.35B. Requires life-cycle cost analysis (including operating cost for energy) to be performed for selection of design alternatives for new schools and state facilities (1986). The Legislature recognized that the operating costs of a facility over its lifetime can exceed the initial cost of the facility. As defined in this section, life-cycle cost analysis means the initial cost and cost of operation of a major facility over the economic life of the facility. This shall be calculated as the initial cost plus the operation, maintenance, and energy costs over its economic life, reflecting anticipated increases in these costs discounted to present value at the current rate for borrowing public funds, as determined by the state finance committee. The energy costs shall be those projected by the state energy office, and the energy use projections shall be updated at least every two years.

Section 43.19.670 - 43.19.685. Requires energy audits and technical assistance studies of energy-conservation potential at all state-owned and state-leased facilities. Establishes a schedule of implementation for energy-conservation measures (ECMs) throughout the state. This 1980 legislation required the directors of general administration and the energy office to conduct an energy audit for each state-owned facility. Not later than 12 months after the audit, both directors were charged with coordinating and implementing energy-conservation maintenance and operation procedures for each state building. Both directors were required to prepare and transmit to the governor and the Legislature the results of the energy audits.

Section 43.21F.010 - 43.21F.420. Establishes the state energy office and enumerates responsibilities and roles. Amended in the 1991 legislative session to require development of a state energy strategy and the facilitation of conservation and cogeneration development at state facilities and schools. This legislation is discussed in section four of this report.

Section 43.21G. Establishes roles and responsibilities of state agencies and the governor in response to energy-supply emergencies (1975). Most state legislatures designed energy-supply emergency plans in the mid-1970s, following the oil embargo. Possible shortages and supply disruptions affected energy transportation, production, and distribution; emergency plans such as this attempt to protect the public health, safety, and general welfare of state citizens. In an emergency situation, the Washington Legislature placed a high priority on delivery of essential government services, emergency services, public mass transportation systems, food production, provision of water to agriculture, and other areas. This emergency plan attempts to provide the equitable distribution of energy among the geographic areas of the state.

Section 43.41.170 - 43.41.175. Allows the office of financial management to permit state agencies and facilities to retain savings from energy-conservation for several purposes, including additional energy conservation. Requires the state energy office to provide the office of financial management with the data necessary to determine retainable savings (1986). The Legislature recognized in this legislation the state's obligation to operate state buildings efficiently and to implement all cost-effective energy-conservation measures so that citizens are assured that public funds are spent wisely and so that citizens have an example of the savings possible from energy conservation. Facilities or the agencies responsible for them are required to report accurate monthly energy-consumption and cost figures for all fuels to the state energy office on a quarterly basis, including any changes in total space served or facility operations.

Section 43.52A. Establishes two Washington state positions on the Northwest Power Planning Council. The State of Washington agreed to participate in the Pacific Northwest Electric Power and Conservation Planning Council pursuant to the Pacific Northwest Electric Power Planning and Conservation Act (1981). The council focuses on regional power planning. This legislation gave authority to the governor to appoint two Washington residents to the council, with the consent of the senate.

Section 54.16.280. Clarifies the authority of local jurisdictions and municipalities to provide financing (through bonds or other means) for energy-conservation measures in existing structures. This legislation permits public utility districts to engage in energy-conservation grants and loans (1979). Any municipality or local authority is allowed to finance the acquisition and installation of materials and equipment, for compensation or otherwise, for the conservation or more efficient use of energy, if the cost per unit of energy saved or produced by the use of such materials and equipment is less than the cost per unit of energy produced by the next least costly new energy resource which the municipality or local authority could acquire to meet future demand. Loans shall not exceed 10 years.

Section 80.50.010. Establishes an energy-facility site evaluation council (EFSEC) to coordinate the authorities of several state agencies concerning environmental and other siting considerations. Amended in 1990 to place EFSEC under the administrative authority of the energy office. The Washington Legislature recognized the pressing need for increased energy facilities due to population growth, and wanted to ensure that the location and operation of such facilities will produce minimal adverse effects on the environment, ecology of the land and its wildlife, and the ecology of state waters and their aquatic life. It was the intent of this original legislation to provide abundant energy at a reasonable cost, while preserving and protecting the quality of the Washington environment.

Section 82.16.055. Allows for deduction from gross income taxable under state public utility tax those amounts expended by a utility to develop energy conservation, or efficiency (1980). In computing tax under this chapter, there shall be deducted from the gross income an amount equal to the cost of production at the plant for consumption within the state of Washington of: (A) 1) electrical energy produced or generated from cogeneration as defined in

RCW 82.35.020; and 2) electrical energy or gas produced or generated from renewable energy resources such as solar, wind, hydroelectric and geothermal energy, wood, wood wastes, municipal wastes, agricultural products and wastes and end-use waste heat; and (B) those amounts expended to improve consumers' efficiency of energy end-use or to otherwise reduce the use of electrical energy or gas by the consumer.

Section 87.03.017. Clarifies the authority of local jurisdictions and municipalities to provide financing (through bonds or other means) for energy-conservation measures in existing structures. This legislation permits public irrigation districts to engage in energy-conservation grants and loans. Irrigation districts are authorized through this 1981 legislation to perform inspections of residential structures to determine the potential for energy conservation, and to provide a list of businesses that sell and install energy-conservation materials and equipment. Irrigation districts also are allowed to arrange for the installation of the conservation materials and equipment by a private contractor whose bid is acceptable to the owner of the residential structure. Loans shall not exceed 10 years.

Energy-efficiency Programs And Initiatives

State Energy Strategy/Cogeneration Legislation

Possibly the most significant legislation to emerge from the 1991 session—subsequently signed into law by Governor Gardner—is Engrossed Senate-House Bill (ESHB) 5245, which relates to a state energy strategy. The legislation addressed three primary areas.

1) It directed the WSEO and an advisory group to develop a state energy strategy,

2) It established a central role for the WSEO in implementing conservation measures in state facilities and schools, and

3) It formally declared that cogeneration power from state-owned facilities an energy-saving priority.

It includes a section on cogeneration incentives for the state that became effective on July 28, 1991. Washington energy policymakers have identified substantial savings associated with using or selling waste steam for electricity production; the new legislation allows the state to profit from identifying and/or building cogeneration facilities.

Cogeneration, as used in Washington legislation, means the sequential generation of two or more forms of energy from a common fuel or energy source. For example, steam heat (usually not captured) released from a boiler at a state agency can be used to run a turbine, which in turn generates electricity that can be sold to a utility. The legislation clearly indicates the purpose of the legislation is not to put the state into the electricity retailing market.

The law mandates that each state agency and school district implement cost-effective, energy-efficiency improvements to minimize energy consumption and related environmental impacts and operating costs. The intent of this legislation is similar to that of other states. The WSEO was appropriated $292,000 over two years to direct and oversee an advisory committee charged with developing a comprehensive state energy strategy.[19] The WSEO will provide a progress report to the Legislature in 1992, with final recommendations and a report due December 1, 1992.

The state energy strategy must consider all forms of energy and identify measures to help maintain adequate, reliable, secure, economically and environmentally acceptable supplies; and identify and quantify, to the extent possible, the costs and benefits of energy alternatives, including direct economic costs and benefits, environmental costs and benefits, and the costs of inadequate or unreliable energy supplies. Over the next year, the committee will attempt to develop a framework in which public decisions and actions that affect energy supply and use can be analyzed comprehensively, to make policymakers aware of the energy costs associated with all public projects.

The collaborative approach used by Washington for the makeup of its committee is worthy of mention. The Legislature realized that a diverse group of people was needed to draft a state energy strategy if it were to be supported and implemented. The legislation mandates that the committee be comprised of 20 members representing different regions of the state, including 15 citizen members appointed by the governor. These appointments are to include at least one member from each of the following: an investor-owned electric utility, an investor-owned gas utility, a natural gas pipeline company, a petroleum product company, a public utility district, a municipally-owned electric utility, a commercial energy user, an industrial energy user, an agricultural energy user, the Association of Washington Cities, the Association of Washington Counties, two people from civic organizations, and two people from environmental organizations. In addition, the committee is composed of other related state agency representatives, one legislator from the house and one from the senate.

Another section of the legislation mandates that the WSEO coordinate with local utilities to develop energy-conservation programs. Once a state agency has identified energy-saving measures, the agency is allowed to "sell" its savings to Washington utilities. Potential benefits from energy-efficiency projects at public facilities include savings in the form of reduced energy costs; revenues from lease payments, sales of energy, or energy savings; and avoided capital costs. Washington utilities are allowed to credit energy bills, give low-interest loans, rebates, or payment per unit of energy saved to the state agencies as compensation for the energy saved by the state. State agencies are allowed to retain all net savings in the form of reduced energy costs, and one-half of all net revenues from any transaction with a utility, the BPA, or other entity.

The net savings are to be retained by the local administrative body responsible for the public facility. The remainder of the savings (50 percent) goes into a new energy-efficiency services account established in the state treasury as part of this legislation. Expenditures from the account can be used only for the state energy office to provide energy-efficiency services to state agencies and school districts, including review of life-cycle cost analyses; and transfer by the Legislature to the state general fund.

An energy-efficiency construction account, also established in the state treasury through this legislation, is to be funded by general obligation and revenue bond proceeds appropriated by the Legislature. Funds from this account are to be used for construction of energy-efficiency projects, including evaluation and verification of benefits, project construction and project administration. Through the legislation, the WSEO is allowed to finance energy conservation in school districts and state agencies.

Life-cycle Cost Analysis. Section 15 of the legislation provides guidelines for life-cycle cost analysis. The WSEO is charged to develop and issue guidelines to promote and define a procedure for selecting low life-cycle cost alternatives. Life-cycle cost analysis measures the initial cost and operating costs for a major facility (building) over its economic life. It is often calculated as the initial cost plus the operating, maintenance, and energy costs over the facility's economic life, reflecting anticipated increases in these costs discounted to present value at the current rate for borrowing public funds.

At a minimum, the guidelines must identify energy components and system alternatives—including energy systems and cogeneration applications—prior to the energy-consumption analysis, and must address energy considerations during the planning phase of a project, among many other provisions.

Cogeneration Benefits. The cogeneration section of this law is designed to increase Washington's energy efficiency by pursuing cost-effective opportunities for cogeneration in existing and new state facilities. The WSEO is charged with notifying all state agencies of the benefits, financing alternatives, and possible applications of cogeneration techniques. Washington utilities are given the first option to participate in any cogeneration opportunities WSEO identifies. WSEO must notify the utility of its intent to build a cogeneration facility and must allow the utility to participate in the feasibility analysis. So long as the feasibility study shows the cogeneration project to be in line with the state energy strategy and the project is cost-effective, the project will probably be approved. Legislators in

Washington recognize the local utility's expertise, and have allowed WSEO representatives to negotiate directly with the local utility for the purpose of entering into a sole-source contract to develop, own, and/or operate a cogeneration facility. If the project is determined to be cost-effective, the state will pursue ownership of the cogeneration facility. If the numbers are less than optimal, the state will not pursue full ownership rights.

Washington's state government has identified the potential for cogeneration projects and increased energy efficiency; both bring revenue to the state treasury. Compensation to the state is allowed from sales of electricity or thermal energy to utilities, lease of state properties, and value of thermal energy provided to the facility, so the state benefits from selling excess steam to produce electricity even if it decides not to own the cogeneration facility. The state is prohibited from seeking regulatory advantage or using its influence to obtain process advantages as a cogenerator of electricity.

The Institutional Buildings Program[20]

From 1982 to 1987, the BPA used an Institutional Buildings Program (IBP) to implement conservation measures in schools, hospitals, governments, and special districts throughout its four-state service territory. Funded and administered by BPA, the program provided incentive payments to eligible institutions for cost-effective conservation measures. The four states in BPA's service region delivered the program to institutions in their states. BPA developed the guidelines for implementing the program and actively monitored program progress.

Regionwide, conservation measures for 348 institutions were funded at a total cost to BPA of $34 million. The WSEO administered the IBP in the state, where the IBP served 642 buildings and helped finance 1,050 energy improvements at a cost to BPA of about $15 million; 121 institutions participated in the IBP program. WSEO estimates the annual cost savings of the program in 1990 dollars is $3.9 million.[21] Actual savings did not match expected savings; analysts discovered that less energy was saved than predicted—between 69 percent and 82 percent of the engineering estimates. Lighting measures comprise 50 percent of Washington's energy conservation measures implemented through the IBP program. Appendix 9 contains the ECM breakdown of Washington's IBP program.

State facilities, school districts and local governments invested heavily in complex and expensive energy-management and control systems over the last decade. Because of expensive and changing technologies that often require system upgrades, the WSEO offers workshops that focus on formation of energy-management control systems (EMCs) user groups. The workshops help users make better economic decisions when facing an upgrade or replacing an antiquated system, wield economic leverage in purchasing systems, share new ideas, and learn about new energy-saving systems.

Participation in the IBP was a multi-step process. After an institution was accepted into the program, an energy audit by a trained auditor was required for all buildings. All operation and maintenance measures identified in the audit had to be completed by the institution. After the audit, the institution signed a contract with WSEO to have a technical assistance survey (TAS) performed for its facility. By signing this contract, the institution agreed to install enough eligible conservation measures so that the total cost of the project was at least 50 percent of the cost of the TAS. The TAS was performed by an engineering or architecture firm selected by the institution. The purpose of the TAS was to:

1) Characterize the existing building and its energy consumption,

2) Note if the operation and maintenance procedures recommended in the audit had been completed,

3) Use a computer model to estimate the existing building's energy consumption,

4) Make recommendations for appropriate conservation measures,

5) Predict the energy savings provided by the recommended conservation measures, and

6) Estimate the cost of the conservation measures.

WSEO reviewed and approved each TAS and all measures eligible for funding. The institution could reject some measures that did not meet the criteria for selection. After the measures were selected, any necessary design work was performed. In some cases, the design process identified additional inappropriate measures. WSEO approval was required for the final design documents. Cost bids were then solicited from contractors to install the ECMs.

Once the contractor was selected, installation of the measures began. In some cases, installation was divided into several phases, with inspections by WSEO after each phase was completed; partial payments were made after a successful inspection. The final site inspection by WSEO and final project approval completed the project. The actual process for final payment to the institution varied. In some cases, WSEO made the final payment to the institution and BPA audited these payments on a regular basis. Later in the program, however, BPA approval on a project-by-project basis was required before final payment was made. After project completion, the institution was required to provide energy-consumption data to WSEO.

248

Puget Sound Telecommuting Demonstration Program

Telecommuting programs, which have been used formally since the mid-1980s, allow employees to work at home, instead of at the traditional work-place, for a portion of the week. The ultimate goal of telecommuting is to provide an alternative to the traditional office environment with no reduction in productivity. The employee usually will have a computer at home, and companies such as Northern Telecom and AT&T have designed a system that allows phone calls to be patched directly to the employee's home, allowing the employee to answer as if actually in the office. It is estimated that 9 million people in the United States telecommute at least on a part-time basis.[22]

Benefits said to result from telecommuting include:

- Less air pollution due to fewer vehicle miles traveled,

- Better employee morale due to work-hour flexibility,

- Less commuter-related stress,

- Less vehicular congestion on highways,

- Reduced highway-maintenance costs,

- Improved job productivity, resulting in increased efficiency,

- Decreased employee turnover,

- Reduced overhead (office space and parking),

- Less sick time taken,

- Increased economic development (especially for high technology manufacturers involved in telecommunications), and

- Greater responsibility and autonomy.

Washington has run a successful telecommuting demonstration program since 1990 in the Puget Sound area. Twenty-three public- and private-sector employers currently participate in the pilot program, with an average of 12 to 15 telecommuters each. Congestion on Puget Sound's Interstate-5 corridor—the sixth worst congestion in the nation—is estimated to cost Seattle citizens almost $1 billion annually.[23] While most telecommuting programs are too new to provide conclusive evidence that the concept *does* reduce traffic congestion, energy consumption, and air pollution, WSEO says that a formal program, if implemented in the Puget Sound area, would save $7 million in annual fuel costs. Preliminary data from the 240 participants indicated estimated annual fuel savings of $25,000 for the first year of the project.

Many organizations and businesses, including the state government, have set up "telework" stations as part of a comprehensive telecommuting strategy. Telework stations are satellite offices, often shared by separate businesses, where groups of employees who live substantial distances from their regular offices can work at an office closer to home. It usually requires 25 or more employees living within the same geographic area to warrant an employer's investment in a telework station. The state government has used a telework station in Seattle since March 1991. Most state government offices are located in Olympia, and many state employees commuted 120 miles round trip daily from Seattle.

The telework center provides each employee with a professional work station, including a computer linked to a local area network (LAN). The employee is connected to the main office through advanced telecommunication products. Support equipment includes a FAX machine, copier, and laser printer. A conference room seating 10 to 12 people is also available for employees. If the telework center proves to be as successful as projections indicate, the center will continue as a state facility serving state agencies in the Puget Sound region.

The State Building Energy-management Program

To reduce the costs of state government and provide an example of energy efficiency to the citizens of the state, Washington legislators passed RCW 43.19.668-680, which authorizes the state to reduce energy consumed in the operation of state buildings and facilities.

The program has six stages: conduct an energy-consumption walk-through survey, implement maintenance and operations recommendations, conduct technical assistance studies, obtain funding for program activities, implement energy-conservation measures, and monitor and report on the progress of the program.

A January 1991 report documents current cost savings at $6.1 million per year.[24] Additional annual savings of $1.5 million per year are attributable to direct purchase of natural gas. Accumulated conservation savings from June 1981 to June 1990 are $38.5 million. Conservation savings will accrue at $8 million to $10 million per year during the 1991-1993 biennium due to the conservation measures to reduce carbon dioxide emissions by 200,000 tons per year.

Energy savings under this program have reached 14 percent and are projected to reach 28 percent. Savings were originally estimated at 33 percent; however, some projects were not started because of the proximity to asbestos, and the decline in fossil fuel prices kept some projects from being cost-effective.

Funding for 1989-1991 was:

Source	Amount
Approved 1989-1991 capital budget	6,005,000
Federal grant funding	150,000
Utilities	250,000
Lease-purchase financing	350,000
Total 1989-1991 biennium	$6,755,000

The majority of conservation financing will come from the state capital budget; however, these projects will be primarily repair or remodel of existing facilities. Third-party financing of energy-conservation projects will be used only for interim financing until permanent financing can be obtained.

The Washington State Energy Code (The Residential Building Code)[25]

During the 1985 session, the Washington Legislature passed SHB 1114 (chapter 19.27A RCW), establishing new energy-efficient building standards for residential structures and directing the State Building Code Council (SBCC) to implement an energy code consistent with these standards. The legislation called for two levels of envelope efficiencies for new buildings, one for electric space heating and the other for all other space heating fuels. From the new energy-efficient building standards, the SBCC drafted the 1986 WSEC.

The WSEC was enforced in all Washington counties, cities, and towns, except for local governments that enforce the model conservation standards (MCS). Due to 1990 legislation (ESHB 2198, chapter 19.27 ARCW), as of July 1, 1991, the WSEC achieves 100 percent of the electricity savings of the MCS. In 1987, the WSEC achieved only 60 percent of the electricity savings of the MCS. The building component insulation levels originally prescribed by the WSEC and the MCS are presented in the appendix 9.

Model Conservation Standards

In its 1983 plan, the Northwest Power Planning Council called on member states to adopt model conservation standards for new residential and commercial construction. The council developed these standards under section 4(f) of the 1980 Pacific Northwest Electric Power Planning and Conservation Act. The residential MCS place design limits on the per-square-foot amount of electricity used for space heating. After years of opposition from homebuilders and a lawsuit, the formal MCS were passed as part of ESHB 2198 in 1990. The legislation underwent years of analysis, a study of

BPA's four-state Residential Standards Demonstration Program, and a comparison of the MCS to savings associated with the 1986 WSEC.

Washington policymakers measure the cost-effectiveness of energy codes by evaluating whether the insulation measures save energy at a cost equal to, or less than, the cost of energy from alternative sources (a new nuclear plant or hydro project, for example). The cost of obtaining electricity resources through insulation measures in new electrically heated homes is only \$.02 to \$.03 per kWh, or half of the \$.04 to \$.05 per kWh estimated for new power plants in Washington state. This means that rates will increase more slowly for all Washington citizens over time if new electrically heated homes are well insulated through the MCS.[26] The standards adopted by the council and the Legislature can reduce space heating requirements by up to two-thirds.

Transportation Demand Management Legislation

Washington has a comprehensive transportation demand management (TDM) strategy in place to reduce energy consumption in the transportation sector. TDM programs are in place in many states; however, in 1991 Washington passed legislation (SSHB 1671) that requires eight of the state's largest counties to adopt TDM ordinances by October 1992, and specifies that major public and private employers—those with more than 100 employees at one work site—within these eight counties must implement TDM programs by October 1993. The legislation establishes the goals of reducing commuter trips and vehicle miles traveled by 15 percent by 1995, 25 percent by 1997, and 35 percent by 1999.

Because some employers already promote alternatives to single occupant vehicles, the new legislation gives full credit for these trip reductions. The legislation also uses comparisons among peers in the geographic area to evaluate performance and requires good-faith efforts to carry out effective programs. Violators of the law can be charged with civil penalties.

TDM programs focus on work-related commuting, since traffic congestion is the heaviest on weekdays when people are traveling to and from work. TDM programs also attempt to move more people in fewer vehicles while promoting alternatives to single-occupant vehicles, such as high-occupancy vehicle (HOV) lanes, car and van pools, biking, walking, telecommuting, and alternative work schedules. Washington's TDM program will help the state meet federal air-quality standards by reducing motor vehicle emissions and will save the state an estimated \$53 million in annual fuel costs, while reducing commuter stress and improving employee morale, says the WSEO.

The Energy-efficient Mortgage Program[27]

The WSEO manages a program designed to advance energy-efficient mortgages (EEMs), which allow buyers to qualify for a higher mortgage payment due to lower monthly energy costs that result from owning an energy-efficient home. If a home is built to the new WSEC, it is eligible for an EEM.

Most Washington bankers will finance an energy-efficient mortgage if the local building department or utility has inspected the home and found that it meets high energy-efficiency criteria. Banks usually allow a monthly loan payment that is no higher than 28 percent of gross income. Because energy-efficiency measures installed in a new home will typically reduce heating and cooling bills at least one-third, a buyer has more monthly income, which qualifies for a stretch on the debt-to-income ratio.

The EEM program is part of a four-state effort to educate the public about loan advantages when buying new, energy-efficient homes. Washington uses a community approach—local task forces comprised of lenders, appraisers, realtors, builders, and utility and local government representatives meet to determine how to make energy-efficient mortgages a reality in their specific area.

The program provides training and educational materials to all the interest groups, and assistance to appraisers and builders in filling out technical forms required by lenders to document energy efficiency. The program has operated for only two years in two pilot counties.

The WSEO also spearheads a four-state communication and coordination effort on behalf of BPA. The states share what has been successful or unsuccessful, trade program materials, and jointly address barriers—such as issues concerning secondary market policies—at the regional and national levels. One important effort by this regional program is to promote the use of a uniform energy label in all new energy-efficient, stick-built, and manufactured homes. The label, placed inside the electric panel or wired to an inside pipe, describes the home's energy package and documents the presence of a mechanical ventilation system, if one has been installed.

Appraisers, lenders, realtors, and builders can use the energy label to verify, document, and advertise the home's efficiency. Code officials and utility inspectors can use it to certify compliance with the energy code or utility standards. It educates the homebuyer and serves as proof of the home's higher value when it is resold. Properly installed, it is permanent, easy to find, and easy to understand.

Here is how an energy-efficient mortgage works. In this example, a Washington resident can afford a home worth $5,100 more—a value significantly greater than the cost of the energy-efficient features.

Facts:
Family income of $2,500 per month
Down payment of 10 percent
Mortgage rate of 10.5 percent
30-year term

Standard Home:
Maximum monthly payment at 28 percent debt-income ratio	$700
Maximum mortgage amount at 90 percent of appraised value	$66,700
Maximum house price mortgage plus 10 percent down	$74,100

Energy-Efficient Home:
Maximum monthly payment at 30 percent debt-to-income ratio	$750
Maximum mortgage amount at 90 percent of appraised value	$71,300
Maximum house price mortgage plus 10 percent	$79,200

Integrated Resource Planning

As part of the Northwest Power Planning Council, Washington uses a regional approach to IRP. Nowhere else in the United States is the concept of regional electricity planning more advanced.[28] Energy efficiency is a common term for northwestern utilities, especially in Washington. The passage of the Pacific Northwest Electric Power Planning and Conservation Act of 1980 (PL-96-501) authorized Idaho, Montana, Oregon, and Washington to collectively plan for and protect certain resources. Each state subsequently passed enabling legislation and the Northwest Power Planning Council was formed.

In 1981, legislation was passed requiring least-cost planning and demand-side management for municipal utilities. Least-cost planning rules for other utilities have been in use since 1987, when the first least-cost plan was prepared by a Washington utility; energy efficiency has since become increasingly important. The Utilities and Transportation Commission (UTC) regulates the state's investor-owned utilities, and has traditionally relied on heavy public involvement and an emphasis on energy efficiency, LCP, and IRP.

In 1987, an LCP rule (WAC 480-100-251) was enacted in Washington that requires electric companies to file biennial plans with the UTC. The commission determines whether the plans meet the minimum requirements of the rule. The rule requires that all resources (supply and demand side) be evaluated on a consistent basis. No formal cost-benefit test has been established. Most electric and gas utilities are using either the total resource cost test or the revenue requirements test.

A statute enacted in 1980 (RCW 80.28.024) provided rate incentives to promote conservation and the use of renewable resources. In the late 1970s, the Washington Legislature gave IOUs a 2 percent higher rate of return on cost-effective investments in energy efficiency. Conservation investments included in the rate base earn a 2 percent higher equity return than the authorized return for other investments. Recent legislation extended the 2 percent return for certain residential construction programs, as well as for conservation programs where priority was given to elderly and low-income customers.

Less than 50 percent of electricity sales in Washington are made by IOUs regulated by the UTC. Remaining sales are by publicly-owned utilities, subject to local regulation and meeting much of their load through purchases from BPA.

All utilities in Washington are dependent on hydropower to some extent, either through reliance on their own hydro resources or purchases from the BPA or other publicly owned utilities. Four companies provide almost all the retail gas in the state. As part of a regulatory experiment, Washington's largest investor-owned utility has decoupled profits from sales since April 1991. A system of rewards and penalties for meeting, exceeding, or missing DSM goals is in place. All IOUs participate every two years in LCP procedures.[29]

An all-resource bidding procedure—considering the environmental impacts of bidding projects—is required for electric utilities by a rule enacted in 1989 (chapter 480-107 WAC). Electric companies must file short-term avoided costs annually. Long-term avoided costs must be filed with RFPs. No rule currently establishes a formal methodology for calculating avoided costs.

The NPPC has included 100 MW of conservation voltage regulation power in its analysis of resources for a new power plan. Conservation voltage regulation is a set of measures and operating strategies designed to provide electricity service at the lowest practicable voltage level.

The NPPC has integrated the potential of renewables (solar, hydro, wind, geothermal, biomass) into its existing conservation supply curves. The council anticipates that this could add approximately 100 MW of energy to the supply curve at an average cost of $0.9 per kWh.

Bonneville Power Administration's *1990 Loads and Resources Study* (also known as the *White Book*) calls for additional supplies to meet BPA's growing demands in the next decade. BPA has 30 hydro dams in the Columbia River Basin and two nuclear power plants. The *White Book* forecasts a demand for power of about 9,800 MW by 2011; BPA's conservation and generating resources will need to grow by about 1,300 MW.[30]

Under BPA's *White Book* program, competitive bid solicitations are being issued for 100 MW of new energy resources. The *White Book* calls for 200 MW of conservation resources by 1997.

BPA has developed an energy-savings transfer program that uses innovative methods to quantify saved energy accurately so that saved power can be transferred from one utility to another. The most unique concept to result from a saved energy measurement and transfer is that energy savings from utility conservation activities become a marketable commodity with market characteristics. This program began in 1988.

BPA, in conjunction with its retail utility customers and other interested parties, conducted an assessment of mechanisms for resource acquisition. BPA is examining how retail utilities will participate in resource development, what role independent power producers should play, and the effect each approach will have on the quality and availability of resources and BPA staffing requirements. Through this process, BPA developed several potential demand-side approaches to resource acquisition.

The Northwest Power Planning Council released its Draft 1991 Power Plan in December 1990, after nearly two years of work.[31] The council considered three factors when designing the 20-year plan: direct costs to build and operate a resource, environmental impacts, and the risks of under- or over-building. The plan calls for action, since the region has exhausted its electricity surplus.

The draft plan has a four-part strategy:

1) Buy all the low-cost, environmentally benign resources currently available;

2) Build an inventory of reliable resources that can be brought on-line quickly if needed;

3) Confirm other resources that are promising or that still pose significant questions (research and resolve legal, environmental and public acceptance issues prior to developing resources); and

4) Remove regulatory barriers that cause utilities to avoid investments in energy efficiency (decoupling profits) and develop incentives for utilities to engage in energy-saving activities.

Conservation is a key factor in the council's resource portfolio for meeting future electrical energy needs. In its 1991 draft plan, the council identified nearly 3,200 average MW of technical conservation in a high-demand forecast, available at an average cost of about $0.04 per kWh. This would be enough energy to replace the output of about six large coal plants, at less than half of the cost.[32] Much of the energy saved on the demand-side is projected to result from improved water heating and improved refrigerator and clothes dryer design.

Table 28
Washington Matrix

WASHINGTON Energy Program	Cost per kWh (cents/kWh)	Annual Savings ($)	Implementation	Environmental Benefits	Start-Up Costs	Fiscal Fairness (+,-)	Probability of Success	Benefit to Cost Ratio (Benefit/Cost)	Limitations or Special Circumstances	Confidence
Institutional Buildings Program (IBP)	3.9/Kwh (1)	$3.9 million M	Moderate (2)	Medium (3)	High (4)	(-) (5)	Medium (6)	1.33:1-M (7)	Time consuming (8)	High (9)
Puget Sound Telecommuting Project	Not available	$25,000 E (10)	Easy (11)	High (12)	Medium (13)	(+) (14)	High (15)	N/A (16)	Infant stage (17)	Medium (18)
State Building Energy Management Program	Not available	$6.1 million M	Difficult (19)	Medium (20)	Medium (21)	(+) (22)	High (23)	1.63:1-M	Private sector influence? (24)	High (25)
Residential Building Energy Code (RCW 19.27)	2.4/Kwh	$36.5 million M	Difficult (26)	Medium (27)	High (28)	(-) (29)	High (30)	1.16:1-M	Legislative support is crucial (31)	High (32)

Matrix Notes

These notes correspond to the numbers in the matrix (**table 28**).

1) All cost per kWh and annual savings figures are stated in constant 1990 dollars.

2) The moderate ranking for implementation of the Institutional Buildings Program is the result of the numerous and diverse institutions with which the agency must work while managing this program. For example, over five fiscal years, the IBP financed 1,050 energy conservation measures in 642 buildings owned by 121 institutions. The implementing agency deals with governments, general hospitals, secondary schools, elementary schools, colleges, and special districts. All these bureaucracies have different management requirements and procedures with respect to energy expenses.

3) Environmental benefits are medium. Additional electricity generation is avoided over the long run.

4) Start-up costs are more than $500,000. Lighting measures (lamp/fixture replacement, lighting redesign, automatic lighting) make up almost one-half of the installed ECMs, and are expensive initial costs. Energy management systems, insulation, and HVAC improvements make up most of the additional improvement expenses. Training costs and engineering estimates add almost 10 percent each to the total costs of ECMs.

5) Fiscal fairness rates a "-", since the people who pay for the program do not necessarily benefit from the program. The audience is relatively narrow.

6) Due to the diverse interests and the multidisciplinary approach needed to achieve success (mentioned in 2 above), probability of success is a "medium" for other states.

7) All benefit-to-cost ratios included in this matrix are measured from actual savings that resulted from this program.

8) This program can take significant time to set up and manage. Most state policymakers would believe the benefits are worth the investment; however, it can take more than one year to organize and implement the program.

9) Program evaluation information was available for this evaluation.

10) This annual savings figure is for gasoline savings to the 240 participants. No other savings (time, environmental cost savings) are included. This is a conservative estimate of the project's total dollar savings.

11) Telecommuting projects for state governments are easy to set up and manage. With the exception of finding commercial space for a telework center, very few components of a state government telecommuting project take much time to arrange.

12) Environmental benefits are immediate and high, in the form of reduced automobile emissions.

13) Start-up costs are close to $250,000.

14) Fiscal benefits are theoretically shared by all state taxpayers, since the project lowers the cost of running state government.

15) Telecommuting projects are likely to succeed in other state governments. Similar projects have achieved success in California and Arizona. European government telecommuting projects have shown positive results, as well.

16) No benefit-to-cost ratio is calculated for this pilot project.

17) Telecommuting projects in the United States are relatively new. This pilot project is also in its trial stage. Nationally, voluntary support from the private sector for telecommuting projects has not been strong.

18) Project information and interviews were used for this evaluation, but no formal program evaluations were ready for review at publication.

19) Implementation of this program is difficult. The program has six stages that require technical expertise: conducting energy consumption surveys, implementing maintenance and operations procedures, conducting technical assistance studies, budgeting, implementing energy-conservation measures and monitoring progress of the installed measures.

20) Additional electricity production is avoided over the long run.

21) Start-up costs are less than $500,000.

22) Fiscal benefits are theoretically shared by all state taxpayers, since the program lowers the cost of running state government.

23) Probability of success is high. Other states have similar programs in place.

24) The program's primary objective is to reduce state government expenditures for electricity and fossil fuels without compromising existing levels of service. A secondary objective was expected

stimulation of conservation efforts by private businesses and local governments. How successful the program was in influencing private sector and local government energy decisions is difficult to estimate.

25) Program evaluations were available for this analysis.

26) Implementation of any residential building energy code on a statewide basis is difficult due to the number of local governments, realtors, home builders and other affected parties involved.

27) Additional electricity generation is avoided over the long run.

28) Start-up costs are more than $500,000.

29) Some residential homeowners are excluded from sharing program benefits due to the age and/or structure of their existing home. In addition, existing tenants in older multifamily dwellings are removed from program benefits.

30) Residential building energy codes produce indisputable benefits. Long-term probability of success is high.

31) Without legislation and legislative support, voluntary state residential building energy codes lack compliance. While the implementation of a similar code is difficult in the short-term for states starting from scratch, the long-term benefits and energy savings are high.

32) Formal program evaluations were available for this analysis.

Future Direction—Legislative Priorities

1) In 1992, many legislators will spend time designing the state energy strategy that was mandated through legislation in 1991.

2) Legislators are expected to have the results of the Puget Sound Telecommuting Demonstration Program in 1992, and the state may legislate action in this area.

3) Legislators will assess the Northwest Power Planning Council's 1991 Power Plan, which was released in late 1991.

4) Cogeneration studies and applications are another area that state policymakers will address in the coming year.

5) The Legislature will review incentives for renewable energy and conservation.

6) State-owned vehicles are targeted for efficiency improvements in 1992. There is legislative interest in natural gas vehicle incentives.

7) There is legislative concern about electromagnetic fields.

8) Washington legislators are interested in Oregon's quantification of externalities resulting from electric and natural gas power production. The Oregon PUC opened a docket on externalities in November 1991, and the PUC order is due in 1992.

9) Some Washington legislators will monitor the progress of the Governor's Task Force on Externalities. The WSEO is responsible for seeing that the task force makes recommendations by the end of 1992 on how to handle this divisive issue.

Chapter Notes

1   *Washington Energy Strategy*, draft prepared for the WSEO by the Washington Energy Strategy Advisory Committee, November 1991, 1A.

2   WSEO, *1991 Biennial Energy Report: Issues and Analyses for Washington's Legislature*, January 1991, 26.

3   *Washington's Energy Strategy*, 2C.

4   1991 *Biennial Energy Report*, 26.

5   Richard Byers, WSEO, November 1991: personal communication.

6   *1991 Biennial Energy Report*, 36.

7   Ibid., 20.

8   WSEO, *1987 Biennial Energy Report: Issues and Analyses for Washington's Legislature*, January 1988, 31.

9   Ibid., 28.

10   Ibid., 31.

11   *Washington's Energy Strategy*, 1B.

12   *1991 Biennial Energy Report*, 40.

13   Michael Farley, WSEO, July 1991: personal communication.

14   *1991 Biennial Energy Report*, 6.

15   Ibid., 30.

16   *Washington's Energy Strategy*, 3C.

17   Ibid., 4D.

18   Ibid., 4C.

19   Richard Byers.

20   Some of the information used in this analysis was obtained from two documents prepared by the WSEO: *Institutional Buildings Program in Washington State: Case Studies of Seven Participating Institutions* (April 1989), and *The Institutional Buildings Program in Washington State: An Analysis of Program Impacts* (December 1987).

21   Richard Byers, December 1991.

22   WSEO, *Telecommuting Fact Sheet*, 1991.

23   WSEO, *Transportation Demand Management*, 1991, 1.

24   Some of the information used in this analysis was obtained from four documents published by the Washington State Department of General Administration:
Energy Conservation in State Facilities: Progress Report to the House Energy and Utilities Committee (January 1991),
Energy Conservation in State Facilities: Progress Report to the House Energy and Utilities Committee (January 1989),
Energy Conservation Program Summary Report for State-Owned Facilities in Washington State (1985-87 Biennium),
Energy Conservation Program Summary Report for State-Owned Facilities in Washington State (1981-83 Biennium).

25   Some of the information used in this analysis was obtained from two documents prepared by the WSEO:
Cost Effectiveness of Residential Building Energy Codes: Results of the University of Washington Component Test and the Residential Standard Demonstration Program (December 1989) and
Your Guide to the 1991 Washington State Energy Code (1991).

26   WSEO, *Cost Effectiveness of Residential Building Energy Codes: Results of the University of Washington Component Test and the Residential Standard Demonstration Program*, December 1989, 5-I.

27   WSEO, *Dispatch* 13, no. 5 (September–October 1990): 8.

28   Frank Bishop, National Association of State Energy Officials (NASEO), Washington, D.C., July 1991: personal communication.

29   Deborah Ross, Washington State Utilities and Transportation Commission, November 1991: personal communication.

30   Robbie Robinson, The Center for Applied Research, Denver, July 1991: personal communication.

31   *Draft 1991 Northwest Conservation and Electric Power Plan*, II, Group 4, 7–2.

32   *1991 Biennial Energy Report*, 12–15.

WISCONSIN

Wisconsin citizens spend approximately $7.5 billion annually to import more than 95 percent of the state's energy from out of state.[1] Wood and hydroelectric power are the state's only major indigenous energy resources. Wood waste (refuse dry fuel) from lumber and construction activities also provides a reliable energy source. Nuclear power supplies almost 25 percent of Wisconsin's energy. Because Wisconsin has no fossil energy resources of its own, state energy programs focus on increased energy efficiency and biomass energy. Energy efficiency has been Wisconsin's greatest resource since the 1970s. The state energy office, the Wisconsin Energy Bureau (WEB), estimates that only one-eighth of Wisconsin's biomass energy potential is used, and that wood alone could indefinitely supply 30 percent of the state's non-transportation energy needs.[2]

Wisconsin is known primarily for its energy-efficient building codes in the industrial and commercial sectors, and for its approach to utility planning. State legislators have sponsored effective legislation in both areas for more than a decade. For example, the Wisconsin Legislature created the Rental Weatherization Program in 1979; it requires that rental dwellings be brought up to energy-efficiency codes before the property can be transferred. The state is also known for its farming, especially the dairy industry. Energy-saving programs in the agricultural area are a priority for state policymakers.

As shown in **figure 42**, Wisconsin energy consumption rose between 1973 and 1978, fell substantially until 1983, and has grown slightly since 1983. Officials at WEB explain that the decreasing trend in energy consumption in the

late 1970s and early 1980s was due to a number of reasons, including high
energy prices, energy-saving public information campaigns at both the state
and federal levels, new energy-saving technologies and incentives at the
federal level, CAFE standards, and a supportive governor and Legislature
interested in energy efficiency.[3]

Figure 42

Total Energy Consumption, Wisconsin, 1970-1988

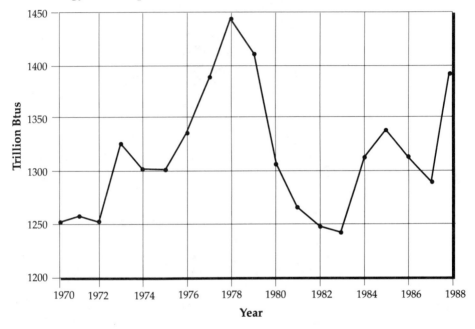

Source: Energy Information Administration, State Energy Data Report, Consumption Estimates, 1960–1988, DOE/EIA-0219, 1988.

WEB representatives say that natural gas delivery problems in 1978
prompted Wisconsin citizens to use less gas, which contributed to the decrease
in energy consumption. Deregulation of the gas industry at the time made it
more profitable for Wisconsin's natural gas suppliers (Texas, Louisiana,
Oklahoma) to sell their gas in-state than to ship it to Wisconsin; as a result,
Wisconsin citizens perceived a natural gas shortage and used less gas.[4]

Increases in energy consumption since 1983 are estimated by WEB to
result from low energy prices, a lack of federal incentives for energy efficiency,
a gradual shift from natural gas to electricity, climatic changes and economic
expansion. Until the early 1980s, when summer days started to reach higher
temperatures of 90 °F to 100 °F, residential air conditioners were an unfamiliar
sight in Wisconsin.[5] New air conditioning units are now responsible for a
growing percentage of coal consumption, since most electric power plants in

the state are coal-powered. Increased coal consumption concerns WEB staff, because it may be responsible for an increase in emissions of SO_2, NOX and SO_2, all thought to be responsible for global climate change. Despite increases in coal use, SO_2 levels in Wisconsin have decreased 50 percent since 1980.[6]

Wisconsin is shifting to coal as an energy source, as shown in 1988 numbers (**figure 43**). Figures from 1991 show the state has increased its use of coal to almost 31 percent of total energy consumption, second only to petroleum, which supplies close to 40 percent of total energy consumption.[7]

Figure 43

Wisconsin Energy Use by Source, 1988 (1,392 Total Trillion Btus)

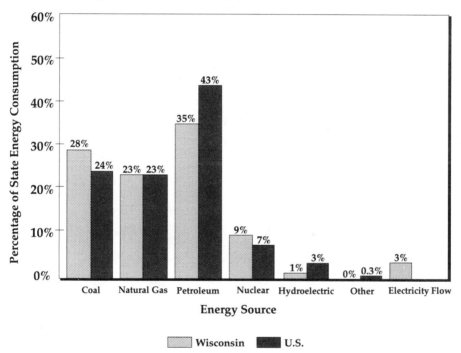

Source: Energy Information Administration, State Energy Data Report, Consumption Estimates, 1960–1988, DOE/EIA-0219, 1988.

The Wisconsin PSC has had the authority to review supply-and-demand options and the environmental costs of power plant construction since 1971. According to WEB personnel, while the PSC was led by environmentalists in the 1970s, recent PSC members have been less aggressive in discovering the environmental costs of electricity production. With respect to energy efficiency, Wisconsin is probably most recognized for its advance planning (AP) process, which is part of the 1975 Wisconsin Power Plant Siting

Law (chapter 196.491, Wisconsin Statutes). This law gave the PSC authority to require from Wisconsin utilities biennial advance plans that include an analysis of the alternatives to the proposed generating and transmission facilities, specifically conservation and renewable energy resources. The law also opened the planning process to include more public involvement in making decisions regarding future energy supply and demand.

Created by the governor in 1973 in response to the OPEC oil embargo, WEB manages many of the successful energy programs in the state and is now part of the state's Department of Administration (DOA). Its mission is to "promote the wise use of energy in Wisconsin by collecting and disseminating energy information, managing energy grant programs, and providing energy policy and technical energy analysis for the governor, legislators, state agencies, and others."[8]

A dozen full-time WEB employees spend the majority of their time monitoring the traditional, federally funded programs—including the Institutional Conservation Program (ICP), the State Energy Conservation Program (SECP), and the Energy Extension Service (EES) Programs—and implementing programs financed by the oil-overcharge funds the state receives; which began after 1983, while many federal programs have been in place since the 1970s.

One project funded by oil-overcharge money is the Wisconsin Center for Demand-Side Research (WCDSR). The center's mission (also described in section four) is to provide analytic support to state agencies and utilities in their energy-conservation efforts in five key areas: technology assessment and development, public policy analysis, economic pricing analysis, customer and market research, and planning and forecasting research. The center received $260,000 in oil-overcharge funds as seed money for initial projects.

Although the center is affiliated with the University of Wisconsin, it is an independent organization that receives contributions from utilities and the private sector. Between 1989 and 1993, utilities have committed between $1 million and $1.8 million per year to the center.[9] Cooperative research projects have been identified that attract diverse organizations to work with the center. Some of these collaborative projects are described in section four.

Energy Consumption by Sector

The Industrial Sector

Wisconsin industry uses natural gas for the majority of its production needs. Wisconsin energy production through renewable resources is minute; however, there have been small advances over the years. In the late 1970s, Packerland Packing in Green Bay, Wisconsin, had the largest industrial solar project in the world—more than 10,000 solar panels. Although still in operation, the project has new plumbing but has reduced the number of solar panels to 5,000. (The remaining 5,000 solar panels are now being sold to the public.) Farming falls in the industrial category, which explains Wisconsin's industrial sector use of more energy than any other state herein—33 percent[*] (**figure 27,** page 135).

The Transportation Sector

The Wisconsin transportation sector comprises 24 percent[**] of total energy consumption (**figure 28,** page 136). While most of the transportation sector's fuel consumption is by vehicles, much of the fuel is used by agriculture. The transportation of feed and livestock contributes to this sector's energy-consumption statistics.

The Residential Sector

Like Washington, Wisconsin is known for its residential energy codes. Residential energy use accounts for 26 percent of total energy use (**figure 29,** page 137). Because of their rural locations and lack of access to natural gas, farmers often do not have access to energy-conservation programs offered by the regulated utilities. To combat the inefficient use of electricity and oil on farms, WEB provided rebates from oil-overcharge funds in the winter of 1989 to farm households that purchased high-efficiency liquid propane, oil, or wood heating systems that met WEB's minimum efficiency standards. This program is discussed in section four.

The Commercial Sector

Wisconsin commercial energy consumption accounts for 17 percent[***] of total energy consumption (**figure 30,** page 138). NCSL was unable to locate any significant Wisconsin energy programs directed at the commercial sector. A business energy fund subsidizes energy-conservation loans to small businesses through private lenders and a new renewable-energy assistance program will provide funding to small and medium-sized businesses to assess the feasibility and installation of renewable energy systems.

Energy-efficiency Statutes

Section 1.12. Directs state agencies to consider energy conservation when making any major decision that would affect energy use. This 1977 legislation was an attempt to alleviate energy shortages.

---

[*]Note: WEB representatives claim the industrial sector uses 28 percent of total consumption.
[**]Note: WEB representatives claim the transportation sector uses 26 percent of the total energy consumed in the state.
[***]Note: WEB representatives claim the commercial sector is responsible for only 8 percent of total Wisconsin energy consumption. The lack of programs aimed at reducing energy consumption in Wisconsin's commercial sector might be due to the sector's relatively low energy consumption.

Section 16.956 (2). Creates a grant program for renewable energy resources and energy-conservation methods. An energy development and demonstration grant program was funded to study renewable energy resources available in Wisconsin, and conservation methods appropriate for the state.

Section 28.11 (3) (k). Permits county boards to establish energy-conservation projects that allow residents to remove wood unsuitable for commercial sale from county forest lands for home heating purposes. No severance tax is required for wood removed from county forest lands for energy-conservation projects.

Section 36.11 (8m). Directs the University of Wisconsin (UW) Board of Regents to require each campus in the university system to develop a transportation plan that results in energy conservation and efficient use of transportation resources. The plans are required to include pedestrian walkways, bikeways, bike routes, bicycle storage racks, car and van pools, and improved mass transit services. The transportation plans shall detail parking-management strategies that provide incentives for the use of mass transit and high-occupancy vehicles.

Section 38.12 (6). Directs vocational education boards to develop transportation plans that result in energy conservation and efficient use of transportation resources. In addition to requiring the same components of the transportation plan mentioned above in section 36.11, this section requires the district board to work with regional planning commissions and local authorities in which the school district is located to design energy-efficient transportation plans.

Section 59.87 (6) (em). Authorizes university extension programs related to energy conservation and renewable energy resource systems. Allows extension agents to conduct programs on energy conservation and renewable energy resource systems, and to provide technical assistance to community agencies, organizations, schools, small businesses, and individuals interested in energy conservation.

Section 66.073 (6m). Directs municipal electric companies to consider energy-conservation measures. Through this legislation, Wisconsin, unlike many states, has required municipal utilities to consider community-based energy-efficiency strategies.

Section 66.073 (10) (b). Requires municipal electric utilities to consider energy-conservation measures as an alternative to additional electric generating facilities. This legislation is an integral part of Wisconsin's IRP process. It requires municipal utilities to consider alternatives to traditional bulk generating facilities when filing their advance plans.

Section 66.299 (5). Requires local government units to award contracts on the basis of life-cycle cost estimates, which may include costs of energy efficiency. The life-cycle cost formula may also include acquisition and conversion, money, transportation, warehousing and distribution, training, operation and maintenance, and disposition or resale.

Section 66.904 (2) (cm) 2. Permits the Milwaukee Metropolitan Sewage District to use life-cycle cost estimates in determining the lowest responsible bid for a contract. Since sewage districts use considerable amounts of energy, this legislation is significant.

Section 77.542 (26m). Exempts from taxation the gross receipts from the sales or use of machinery and equipment used to recover energy from solid waste. Gross receipts from the sale or storage of parts or equipment used for recycling activities, reusing solid waste, recycling solid waste or reducing solid waste are also exempted from taxation through this legislation.

Section 85.063 (3) (a). States the legislative finding that development of urban rail transit systems will enhance citizen welfare by conserving fuel. This legislation attempts ultimately to reduce petroleum dependence in Wisconsin's transportation sector. The legislation also mentions the importance of alternative transportation modes and that air-quality improvements are expected with new rail transit systems.

Section 85.065 (1) (a) 4. Permits local governmental units to apply to the Department of Transportation for a study of the extent to which the location of rail lines in an urban area serves the public interest in conservation of land or energy resources. This legislation embodies principles inherent in other states' comprehensive growth management strategies.

Section 85.08 (4m) (a). States the legislative finding that freight rail service preservation affects energy conservation. It is the intent of this legislation to promote the public good by preserving and improving freight rail service in Wisconsin by assisting local governments to preserve freight rail service in those areas of the state confronted with the possibility of service discontinuation.

Section 85.24 (1). States that the purpose of a ride-sharing assistance program is to promote energy conservation, reduce traffic congestion, improve air quality and enhance the efficient use of existing transportation systems. This legislation resembles other states' comprehensive growth management strategies.

Section 101.02 (15) (j) - 101.08. Concerns the energy efficiency of fluorescent lamp ballasts. Public building rules or regulations are not allowed to increase the maximum energy use (as defined in section 101.08) allowed for a fluorescent lamp ballast, or decrease the minimum energy efficiency required for a fluorescent lamp ballast.

Section 101.122 (2) (a) 1. Requires the Department of Industry, Labor and Human Relations (DILHR) to promulgate minimum energy-efficiency standards for rented dwelling units. The rules that resulted from this legislation require installation of specified energy-conservation measures in rental units. The present value benefits of each energy measure, in terms of saved energy over a five-year period after installation, shall be more than the total present value cost of installing the measures. The rental weatherization program that evolved from this legislation is explained in section four.

Section 101.62 - 101.72. Requires the dwelling code council to recommend rules that provide for energy conservation in the construction and maintenance of dwellings. This legislation requires the council to recommend variances in the rules for different climates and soil conditions.

Section 110.20 (3) (c). Permits the vehicle inspection program of the Department of Transportation to provide information on vehicle fuel efficiency. The inspection and maintenance program may be designed to provide fuel efficiency information; however, it is not required.

Section 134.81 (2). Requires manufacturers of water heaters to attach a warning that a thermostat setting above 125 degrees Fahrenheit may consume unnecessary energy. This 1987 legislation requires that a plainly visible notice inform the owner about unnecessary energy consumption and potential severe burns.

Section 159.05 9 (1). States that solid waste reduction and resource recovery conserve resources and energy. This section states that the maximum solid waste reduction, reuse, recycling, composting, and resource recovery are in the best interest of the state.

Section 159.13 9 (8) (c). Permits a municipality to determine that required use of a recycling facility is in the public interest if required use will conserve energy. The emphasis is on conserving both energy and natural resources.

Section 196.373 (2). Requires public utilities that provide gas or electricity to recommend to their customers that water heater thermostats be set no higher that 125 degrees Fahrenheit. This information must be included at least annually in customer billing statements.

Section 196.374 (1). Requires gas and electric utilities to spend annually at least 0.5 percent of their total annual operating revenues on programs designed to promote energy conservation. The PUC may require a utility to spend an amount that is more or less than 0.5 percent of its annual operating revenues if, after notice and hearing, the PUC finds that the expenditure is in the public interest.

Section 230.215 (1) (a). States the legislative finding that flex-time scheduling often results in a more economical and efficient use of energy. Worker productivity benefits, reduced absenteeism, improved employee morale and reduced use of highways are mentioned in the legislation as benefits from flex-time scheduling.

Section 234.49 (2) (a) 9. a. Authorizes the Wisconsin Housing and Economic Development Authority to specify an interest rate lower that the ordinary current rate for housing rehabilitation loans if a substantial portion of the loan will be used, among other things, for energy-conservation improvements. Energy-conservation loans are treated as long-term improvements in the physical structure of the building.

Energy-efficiency Programs And Initiatives

The Rental Weatherization Program

The Rental Weatherization Program (section 101.122 of the Wisconsin Statutes) created by the Legislature in 1979, directed the DILHR to develop for rental units energy-conservation standards that have a payback of five years or less. Although not fully implemented until 1985, this legislation is still considered a model for other states. One of the major benefits associated with this program is that the financial burden of energy-inefficient rental units is shifted from tenants—who are usually distanced from investments in energy efficiency—to landlords by requiring them to upgrade their buildings at the time of property transfer.

Unless a property transfer is shown to be excluded from this law, a DILHR transfer authorization must accompany the transfer documents for rental property when presented for recording the deed. There are three types of transfer authorizations: a certificate of compliance, which means the property meets the standards; a stipulation, which allows the purchaser one year to meet standards; and a waiver, which allows the purchaser two years to raze the building.

This law prohibits the transfer of rental property unless it is weatherized to save energy. Weatherization improvements required under this legislation include double-glazed windows, self-closing storm doors, weatherstripping, caulking, moisture-control (ventilation improvements), efficient water heaters (insulation and efficiency), low-flow showerheads, air conditioners (must be sealed), and insulation (attic and walls). Proof of weatherization is made by private, state-certified inspectors who are hired by building owners to check properties for compliance. There were 659 certified independent inspectors in 1990, and 49 government agencies that could validate stipulations and waivers. In 1990, the program was run by 1.5 professionals, two paraprofessionals, and 1.5 clerical staff. Expenditures in 1990 were $308,000 and revenues were $195,000. After the first official year of the program in 1985, 1 percent of eligible rental buildings were in compliance. After 1990, 14 percent of the state's buildings were in compliance.[10]

In 1985, 3,337 stipulations were issued, compared to 2,267 certificates and 88 waivers. In 1990, 5,774 stipulations were issued, compared to 1,158 certificates and 130 waivers. Owners with stipulations bring the building into compliance with the law 73 percent of the time. Owners who fail to bring their buildings into compliance with the law are referred to the county district attorneys or the state attorney general's office.

The Farmer's Network Program

The Agricultural Resource Management (ARM) Division of the Wisconsin Department of Agriculture, Trade and Consumer Protection (DATCP) manages the Wisconsin Sustainable Agriculture Program. Sustainable agriculture is designed to enhance environmental quality and provide food, while benefitting society as a whole through the efficient processing and distribution of agricultural products. The non-agricultural public, driven primarily by food safety and environmental concerns, shows a continued interest in sustainable agriculture.

DATCP regulates the farming industry, milk, meat, marketing methods, weights and measures, and other aspects of Wisconsin commerce. ARM is the environmental protection division of the DATCP and, in addition to working on traditional water pollution control, pesticide applications, and soil erosion control issues, is also interested in energy savings associated with farm programs.

Each year, projects are selected by ARM and a citizens' advisory council and scored on three main criteria: environmental benefit, energy savings and farm profitability. The projects usually are funded through oil-overcharge funds. As of June 1990, 77 projects had been funded over three years, at a relatively low cost of $1.5 million. One of the more successful, inexpensive programs to emerge from Wisconsin is the Farmer's Network Program.[11]

Traditionally, this program has had strong legislative support. It is designed to educate farmers about the benefits of collective action versus the traditional individualist approach to farming. Objectives of the program are to develop an on-farm research process that involves farmers in the energy-saving decisions of running the farm business from start to finish. Farming techniques—including controlled grazing, mechanical weed control, diversified crop rotation, and improved nutrient management—are used to lower the total energy required for agricultural products. ARM joins a group of farmers who have expressed an interest in participating in the project. They designate (hire) a coordinator—usually a young farmer/graduate student—to oversee the work of the different farms and evaluate the effectiveness of various energy-saving farming techniques. The coordinator maintains contact with state agencies and is responsible to monitor the farmers' progress. Educational events are usually arranged by the coordinator at the end of the project to transfer the knowledge gained to other Wisconsin farmers. According to ARM representatives, "local control" of the project is appealing to rural farmers who participate in the program.

Wisconsin agricultural representatives say the program is successful because it lets the "farmers do the farming" and lets the graduate student/farmer work on the farming as well as on the energy-saving calculations.[12] Wisconsin officials say farmers are interested in environmental issues and want to reduce energy costs. Since studies show that farmers get most of their information from other farmers, the Farmer's Network Program encourages farmer-to-farmer communication by providing technical assistance to farmers who are interested in sustainable agriculture policies. Farmers trained under this program, say ARM representatives, are more likely to share energy-saving information in the community, while taking leadership roles on local and state planning issues.

Farmers involved in the program participate in farm tours, where energy-saving techniques are exhibited, and are active in field days (social and educational events) and workshops. Two farmers' networks composed of approximately 10 farms each were funded in 1990; 10 were funded in 1991.

The Wisconsin Conservation Corps

The Wisconsin Conservation Corps (WCC) funds energy-efficiency improvements in local government and nonprofit facilities, areas sometimes ignored by the five traditional federal energy programs. The program also installs energy improvements in houses, rehabilitation centers, town halls, hospitals, Indian reservations, and universities.

The WCC is under contract with the WEB as part of a four-year, $2.2 million program. Original funding was provided from oil-overcharge monies. With an annual operating budget of approximately $300,000, the program has a full-time staff of four employees, in addition to six part-time technical staff. The WCC, actually a state agency, is governed by a seven-member volunteer board appointed by the governor. Members serve six-year staggered terms, and are selected to provide regional, environmental, and agricultural representation. The board establishes policy for the WCC, selects WCC projects, and hires the agency's executive director.[13]

The WCC hires unemployed, inner-city youth between the ages of 18 and 25 to install energy-conservation measures in nonprofit facilities, government buildings and, more recently, homes. The WCC generally employs 25 youths. Since 1987, the WCC has employed close to 200, who are paid the higher of the federal or state minimum wage and are allowed to work a maximum of one year in the program. At the end of the year, a $500 cash bonus or a $1,800 tuition voucher for use in the state university system is given to each worker.

The WCC hangs doors; installs fixtures, wiring, and insulation; and handles a number of carpentry-related tasks as well as installing lighting, drywal, and weatherization improvements. Licensed electricians finish the wiring started by the WCC. Most WCC work is in the Milwaukee suburbs, although program managers say work is performed throughout the state. The WCC also provides "life-skills"—such as obtaining a driver's license, writing a resume, and learning first aid—to the inner-city youths, although the primary emphasis of the program is on developing good work habits, working with others, and establishing a good work record. Work crews are comprised of multi-cultural groups.

Since the program's inception in 1987, more than 300 projects have been completed by the WCC. The energy-efficiency improvements must have a simple payback (total cost divided by annual savings) of less than 10 years before the WCC will install them.

Biomass and the Wood Waste Energy Incentive Program[14]

Biomass energy is derived from sources such as wood, agricultural crops, animal wastes, and municipal wastes. Unlike fossil fuels, biomass fuels represent a sustainable and ongoing resource for future use. The WEB considers wood use for fuel economical and environmentally sound. This program was started because the state's secondary wood-using industry (building supply retail stores, lumber companies, log home manufacturers, and wood product manufacturers) generated significant amounts of wood waste that could be used as fuel, and because many smaller businesses that generated wood waste found it difficult to dispose of their waste economically.

Wisconsin's Wood Waste Energy Incentive Program was funded in December 1988, with $325,000 allocated from oil-overcharge funds for the installation, repair, or modification of commercial and industrial wood energy systems. Twenty-four businesses received grants from the program for 26 separate wood energy projects. Nineteen of these projects involved installation of new wood energy systems, two expanded existing systems, and five repaired or modified existing systems.

The state investment in renewable energy (wood) was approximately $1 per million Btu, or 0.25 percent of the average commercial cost of fossil fuels in Wisconsin. Energy expenditures of over $1 million per year remain in the state as a result of these wood energy projects. In addition, the project led to the annual use of over 40,000 tons of wood waste as fuel. A typical grant was under $8,000, with 90 percent of the grants less than $22,000; the largest grant was $53,900. The grants were awarded in three cycles between March 1989 and March 1990.

WEB mailed self-addressed cards to 1,400 Wisconsin wood products businesses to identify firms willing to participate in a detailed wood waste survey. Over 250 responded affirmatively, and 120 subsequently completed a detailed survey form. Through this survey, WEB concluded that between 250 and 300 businesses in the state were generating significant amounts of excess wood waste that could be used as fuel. Rising landfill costs were a major concern to businesses who responded to the survey. Based on these findings, the Legislature's joint committee on finance funded the project ($324,000 in December 1988).

Incentives generally comprised 10 percent to 20 percent of the businesses' total costs. Systems eligible for the program ranged in output from 500,000 Btu/hr to 17 million Btu/hr. The requirements of the project were a simple payback of 10 years or less, and a prescribed economic analysis to document the payback, stamped by a registered professional engineer.

Wood Waste Energy Incentive Program
Summary of 23 Projects

	Total	Average
Wood Waste Produced (tons/yr)	92,700	3,700
Wood Waste Used (tons/yr)	41,300	1,800
System Output (MBtu/hr)	110	4.3
Capacity Factor	NA	43
Annual Use (MBtu)*	323,000	14,000
Incentive Amount ($)	303,800	13,200
Incentive (% project costs)	12	18
State Investment ($/MBtu)	0.94	2.35

* Totals for repair/modification projects not included.

The Waste-to-Energy Program[15]

Waste-to-energy projects are important in Wisconsin and the state has aggressively pursued opportunities in this area. At the end of 1990, for example, Wisconsin signed an agreement with the Wisconsin Power and Light Company that will save the state approximately $150,000 annually in landfill costs by turning fly ash waste into energy. The agreement was negotiated between the state's DOA, the Department of Natural Resources (DNR) and the utility's environmental affairs department. Ten thousand tons of fly ash produced by a heating and cooling plant was formerly disposed in a landfill. Since the fly ash has unburned coal properties, some 30 tons per day are now trucked to a waste-to-energy plant. The state saves landfill costs and has extra space in its landfills as a result. Wisconsin companies are test-burning scrap tire chips, as well.

The Waste-to-Energy (and Recycling) Program is a state-of-the-art, cost-sharing grant program administered by WEB. Grants are available for three types of projects:

1) Waste-to-energy grants will support feasibility studies, testing, and technical assistance for applications in which a waste product will be converted into a waste fuel;

2) Recycling grants can be used to provide information on recycling methods, support feasibility studies, evaluate recycling methods through pilot projects, and support ongoing recycling efforts that need technical assistance,

3) Waste-wood, energy-system grants will help pay construction costs for installation or repair of commercial and industrial boilers that use waste wood as fuel.

The program was designed to help firms, municipalities, nonprofit organizations, and state agencies make knowledgeable decisions about waste-to-energy and recycling options. The program is intended to reduce expenditures on fuels imported to Wisconsin, develop a more diversified energy supply for the state, increase competitiveness of Wisconsin businesses through lower energy costs, and improve environmental quality.

For waste-to-energy and recycling projects, grants will cover only 50 percent of eligible costs, so a 50 percent match is required. Wood energy system grants usually do not exceed 20 percent of the total eligible project costs. Among 11 new projects funded in 1990 in the waste-to-energy area, five were related to producing or burning refuse-derived fuel (RDF), three dealt with the use of methane gas escaping from landfills, one evaluated incineration with energy recovery, one produced a wood fuel from used wood pallets, and one used methane gas produced from industrial waste water. For waste-wood systems, construction of cogeneration facilities and construction of waste-wood space heat systems are typical projects funded for energy savings. Payback periods (simple payback) are usually required to be 10 years or less.

The program began in July 1987, and grant cycles have been announced every six to eight months. More than $3 million in oil-overcharge funds have been allocated by the Legislature and the governor to the program. Most of the money has been awarded to waste-to-energy projects. For example, in 1990 $385,000 was appropriated:

Waste-to-energy	$275,000
Recycling	0
Waste-wood systems	$110,000

The Energy-efficient Furnaces for Farms Program

Replacing a liquid-propane (LP), forced-air furnace that is over 10 years old can save up to 30 percent on heating energy use.[16] Replacing an oil furnace or boiler or an LP boiler can save up to 25 percent. Savings are greater, of course, if the replaced equipment is in poor condition. The farm furnace program is a rebate program administered by WEB designed to help families pay for new, energy-efficient oil, LP, and wood heating systems in their farm homes. It was designed to address the lack of energy awareness in rural areas. Since these areas do not benefit from regulated utility conservation programs, farmers are often unaware of energy-efficiency advances. This program involves a population not ordinarily thought of as "big" energy users. However, old farms usually are energy-inefficient and lack the technology necessary for energy-efficiency gains. Elderly farm households (all members are age 60 or older) are eligible, as are farm households with an annual reported federal income totalling at least $5,000.

Rebates are available for heating systems that meet reasonably high annual fuel utilization efficiencies (AFUEs). Rebates averaged $500 for an oil, LP, or wood boiler; $400 for wood forced-air furnaces; $350 for oil or LP forced-air furnaces; and $250 for wood stoves that met EPA guidelines.

Wisconsin Center for Demand-side Management

Wisconsin's Center for Demand-Side Management is a relatively new, but nationally recognized, organization in the energy-efficiency area. Originally funded from oil-overcharge money, operating funds totalling between $1 million and $2 million annually now come from a number of sources, including utility contributions. The center, governed by an 11-member board of directors, is committed to medium- to long-term research on technology transfer and gas and electric DSM advances specific to Wisconsin.[17]

The center provides analytic support to state agencies and utilities in their energy-conservation efforts in five primary areas: technology assessment and development, public policy analysis, economic pricing analysis, customer and market research, and planning and forecasting research. Affiliated with the UW, the center is an independent, not-for-profit organization.

The center's activities have included development of a statewide database study of high-efficiency appliance market potential, improved commercial cooling technologies, energy-conservation demonstration projects, analysis of direct load control, and study of commercial building energy use.

The Community Energy-efficiency Program[18]

This two-year program, funded with $300,000 in oil-overcharge funds in 1990, is run by the UW extension. It targeted two small Wisconsin communities, Mauston and Iola, for intensive community-based, energy-conservation activities. Outreach workers in each community coordinate all energy-conservation activities and work to demonstrate the positive economic effects of energy conservation on the community as a whole. Activities include neighborhood-wide residential energy audits with individual and group feedback; assistance with contracting for, and financing of, conservation measures; and financial incentives. Similar activities were carried out in the commercial, industrial, and public building sectors. A formal evaluation of this program is currently underway.

Integrated Resource Planning

The WPSC is known for its aggressive position to use demand-side measures as a source of electricity supply. The WPSC actually has had the authority to review supply and demand resource options, environmental costs and alternatives to power plant construction since the early 1970s.[19] Wisconsin environmental groups did not need to actively advocate utility DSM during the 1970s, since the WPSC staff had a strong interest in using conservation sources as a supply option.

Wisconsin's formal long-range planning process began in 1975 with the Power Plant Siting Law (PPSL) (chapter 196.491, Wisconsin Statutes), which required biennial long-range planning and a coordinated multi-agency review. The PPSL opened the planning process to public participation and gave the WPSC authority to require and review long-range utility plans before resources could be committed.[20] The law required Wisconsin utilities to design biennial advance plans that include forecasts of energy and peak demand requirements over the next 20 years, plans for the construction of generating facilities over the next 15 years, and plans for construction of proposed transmission facilities over the next 10 years.

Through the PPSL, the WPSC has the authority to modify or reject the plans submitted by utilities. In past years, plans have been rejected because nuclear plants were included as a supply source, and have been modified because not enough conservation measures were included.

In 1979, the Legislature enacted a bill to require the commission to prepare an alternative eiectric power supply study (AEPSS) to assess the state's potential to meet long-range electricity demand through conservation and renewable sources (chapter 350, section 26r, Wisconsin Statutes). The first AEPSS was published in 1982. In 1983, the commission used the AEPSS and other testimony to support its determination that conservation and load management were necessary to state energy planning. In its May 1983 decision, the commission officially called on utilities to increase their reliance on conservation and renewable energy sources. This established a goal to avoid all new electricity construction until the year 2000, and to require utilities to incorporate energy efficiency as a supply source. The first IRP plan,

Table 29
Wisconsin Matrix

WISCONSIN Energy Program	Cost per kWh (cents/kWh)	Annual Savings ($)	Implementation	Environmental Benefits	Start-Up Costs	Fiscal Fairness (+,-)	Probability of Success	Benefit to Cost Ratio (Benefit/Cost)	Limitations or Special Circumstances	Confidence
Rental Weatherization Program	2.1/Kwh	$4 million E (1)	Difficult (2)	Medium (3)	Medium (4)	)(-) (5)	Medium (6)	2:1-E	A) Stipulation requirement B) Enforcement is difficult (7)	Medium (8)
Farmer's Network Program	Not available	Not available	Moderate (9)	High (10)	Medium (11)	(-) (12)	Medium (13)	Not available	Tracking problems (14)	Medium (15)
Wisconsin Conservation Corps (WCC)	3.7/Kwh	$320,000 E	Difficult (16)	Medium (17)	Medium (18)	(-) (19)	Medium (20)	1.7:1-E	Limited improvements (21)	Medium (22)
Biomass and Wood Waste Energy Incentive Program	Not available	$1.9 million E	Easy (23)	High (24)	Medium (25)	(-) (26)	High (27)	5:1-E	Public opposition to combustion (28)	Medium (29)

designed in 1987, called for more than 1,000 MW of demand-side program options.

The WPSC has also designed a non-combustion credit mechanism to credit electricity power options that do not use combustion. Utilities are required to reduce the technical cost of non-combustion options by 15 percent to account for the environmental benefits provided by the non-combustion options. Today, the commission evaluates demand-side programs using both quantitative (cost-effectiveness, energy and demand savings, net benefits) and qualitative (depth and breadth of DSM adequacy of evaluation, inclusion of new technologies in programs) factors. Wisconsin utilities spent close to $130 million on DSM programs in 1990.[21]

Matrix Notes

These notes correspond to the numbers in the matrix (**table 29**).

1) Estimated annual savings are $4 million.

2) Because of the number of entities involved, implementation of the rental weatherization program is difficult. Any statewide building code requires extensive training of building code inspectors before local governments can implement the codes. Wisconsin's program has 659 independent state-certified inspectors and 49 government agencies that participate in the program.

3) Environmental benefits for the program are medium. Additional electricity generation is avoided over the long run.

4) Start-up costs are medium. In 1990, program expenses were $308,000, and revenues amounted to $195,000.

5) The fiscal fairness rating is a "-". Program benefits are not available to all who pay for them.

6) Probability of success is high if, unlike in Wisconsin, the stipulation requirement is enforced. Minnesota has a comparable program in place, and Massachusetts and Virginia are considering similar programs.

7) Energy-efficiency improvements installed as part of Wisconsin's program must have a payback period of five years or less. Wisconsin's colder climate might lead to greater results than would be achieved in a southern state with warmer temperatures. Longer payback periods might be appropriate for other states. The "stipulation" is a major problem for program managers. Buyers are not made aware of total costs of improvements in the beginning, and tend to lose interest in making improvements after a year. One program manager said the problem could be solved by requiring an

inspection prior to signing the stipulation, so the buyer knows in advance the costs of all required improvements. The cost of improvements is approximately $1,000 for the average Wisconsin rental unit, with a five-year payback. Enforcement is lacking in urban areas. As one Wisconsin official noted, "Crime is more important than whether or not your home meets the code."[22]

8) No formal program evaluation was available for this analysis. Interviews and annual program summaries were used to evaluate this program.

9) Because of the rural nature of the program and the wide variety of individuals participating in the program, implementation deserves a moderate rank. As one Wisconsin energy staff person suggested, "Finding a coordinator and bringing the farmers together is not a problem, but convincing farmers to implement energy-saving techniques can be challenging."[23]

10) Environmental benefits are high and immediate. Fewer passes through the field mean reduced diesel use and reduced nitrogen accumulations. Sustainable agriculture practices generally yield high environmental benefits.

11) Start-up costs for this program are less than $500,000.

12) A "-" rating is earned, since all those who pay (citizens, from oil-overcharge funding) do not personally share the benefits. The program's audience is limited to farmers.

13) Due to the individual nature of farming communities, it is difficult to estimate how successful a similar program might be in another state. Since the program's audience is limited and identifiable, the program deserves a medium ranking.

14) Farmers learn from other farmers, and the field day events sponsored through this program provide an excellent forum for new developments in energy-saving technologies. However, tracking the implementation of specific energy-saving techniques can be a challenge. The coordinator must travel frequently and maintain close contact with program participants if accurate results are to be collected.

15) No formal program evaluations were available for this analysis. Interviews and program summary materials were available for this evaluation.

16) The difficult rating is the result of a number of factors. The WCC is a state agency, and its governing body is a seven-member volunteer

board appointed by the governor, who attempts to provide regional, environmental, and agricultural representation on the board. Members, who serve six-year staggered terms, establish policy and select projects for the WCC. Board members also hire an executive director for the WCC. The implementing agency in Wisconsin is charged with monitoring more than 100 projects per biennium, 11 staff (five program staff and six technical staff leaders) and 25 youth at a time from Wisconsin's inner cities. Another complicating factor is that "sponsors" plan and buy materials for projects. Sponsors can be federal, state, or local agencies; Indian tribes; or nonprofit organizations. Program staff say it is a challenge to hire the unemployed and instill a work-ethic in new employees.[24]

17) Environmental benefits are medium. Since 1987, the project has been responsible for the installation of weatherization, insulation, drywall, lighting, and other conservation measures in more than 300 projects. This number is large enough to be difficult to manage, but not large enough to yield high environmental benefits.

18) Start-up costs can be less than $500,000.

19) Fiscal benefits accrue primarily to projects relatively close to major metropolitan areas. Rural locations are not given similar opportunity to participate in the program benefits. It would not be environmentally or economically sensible, in most cases, to transport inner-city youth to rural locations for the day.

20) Because of the large number of stakeholders involved in the implementation of this program, probability of success is judged to be medium. Since the payback requirement is a reasonable 10 years, and most of the measures installed are proven energy-saving technologies, the probability of success of individual energy-conservation measures is high.

21) More complex and technical energy-saving devices are not installed in many projects simply because workers do not have the necessary skills to install them. Underskilled workers provide a great service by installing less complex measures, but many new energy-saving devices are excluded from projects. For example, since WCC workers are not allowed to remain with the program longer than one year (the purpose of the program is to educate and enable Wisconsin youth, not to make the program a final working place), they do not acquire the special skills necessary to install computer-operated, thermostatically-controlled heating and cooling systems.[25] The program is obviously not designed to provide these skills, and this does limit the total energy savings achieved through WCC projects.

22) Interviews and program summary materials were used for this analysis. No formal program evaluations were provided to NCSL.

23) The audiences for similar programs in other states are easy to identify. Biomass systems are relatively easy to build and operate. Administrative requirements for similar programs are minimal.

24) Most biomass or waste-to-energy programs that reduce landfill volume and produce energy through combustion of waste rank high in immediate environmental benefits.

25) Start-up costs are less than $500,000.

26) Program benefits are targeted to a narrow audience, despite the fact that all citizens pay for the program.

27) Probability of success is high. Other than public concern for combustion and air quality, similar programs have relatively few obstacles to overcome. Rural locations generally are more likely to implement these projects.

28) The economic and environmental benefits of these projects are easy to understand; however, public opposition to combustion of waste (especially in urban locations) can be significant. Cities with air quality problems may have to install more expensive pollution control equipment to limit air emissions. The annual savings figure of $1.9 million did not match program information sent to NCSL.

29) No formal program evaluations were available and, as mentioned above, the annual savings figure did not match the program information sent to NCSL.

Future Direction—Legislative Priorities

Wisconsin legislators will consider the renewal of wind energy tax credits and incentives for renewable energy forms in 1993. Alternative-fuel vehicle legislation will be a priority during the next legislative session, and legislators are interested in natural gas vehicle incentives both for consumers and for the businesses that produce the vehicles. The legislature will also consider waste-to-energy legislation, statewide energy planning, and efficiency standards for state government buildings during the next legislative session. In spring 1992, Wisconsin's governor vetoed innovative legislation approved earlier in the year by the Wisconsin Legislature. Assembly Bill 590 would have required state agencies "and those they regulate" to establish an energy supply hierarchy for future energy needs, using this priority list: conservation, noncombustible renewable energy resources, combustible renewable energy resources, and fossil fuels. Where possible, state agencies were to rely on fossil fuels for future energy needs only as a last resort. This legislation started as a legislative petition during the Persian Gulf War, and ultimately became formal legislation which passed one Wisconsin house unanimously. A similar bill is expected to be introduced in 1993. The Wisconsin Legislature will consider environmental externalities during the next legislative session. In May 1992, the WPSC levied a \$15/ton tax on utility CO_2 emmissions. In an attempt to promote renewable energy electricity production. The WPSC action is likely to affect future state legislative energy decisions in numerous states.

Chapter Notes

1   Wisconsin Energy Bureau, *Annual Report—1989*, March 1990, 12.

2   Don Wichert, WEB, November 1991: personal communication.

3   Ibid.

4   Ibid.

5   Don Wichert, July 1991.

6   Ibid.

7   Ibid.

8   Terry L. Johansen, "An Analysis of Wisconsin's Oil Overcharge Funded Energy Programs," Master's thesis (Madison: University of Wisconsin, 1990), 17.

9   Paul Newman, WPSC, November 1991: personal communication.

10   Information for this analysis was obtained from several sources, including: personal communication with Ergun Somerson, Wisconsin Department of Industry, Labor and Human Relations, November 1991;

     1990 Annual Report, Rental Weatherization Program (Safety and Buildings Division, Wisconsin Department of Industry, Labor and Human Relations, February 1991);

     Wisconsin Administrative Code, Rules of Industry, Labor and Human Relations, Rental Unit Energy Efficiency, chapter ILHR 67; and

     Important Information for Owners and Tenants of Residential Rental Properties: Conservation Through Weatherization (Madison: Rental Weatherization Program, April 1991).

11   Information for this analysis was obtained from several sources, including: personal communication with Michelle M. Miller, Wisconsin Department of Agriculture, June 1991;

     1998 and 1989 Profitability Evaluations, Sustainable Agriculture (Wisconsin Department of Trade and Consumer Protection, 1990); and

     Don Wichert, November 1991: personal communication.

12   Michelle M. Miller, June 1991: personal communication.

13   Information for this analysis was obtained from several sources, including: personal communication with Topf Wells, WCC, November 1991; and the *Wisconsin Conservation Corps Fact Sheet* (WEB, January 1991).

14   Most of the information used for this analysis was obtained from *Wood Waste Energy Incentive Program Summary Report* (WEB, October 1990).

15   Most of the information used for this analysis was obtained from *The Waste-To-Energy and Recycling Program, Program Evaluation* (WEB, July 1990).

16   Some of the information used for this analysis was obtained from Terry L. Johansen, and

   Wisconsin Energy Efficient Furnaces for Farms Program Fact Sheet (WEB, January 1991).

17   Information was obtained from written materials and a speech made by Shel Feldman, Wisconsin Center for Demand-Side Management Research, at "How Energy Efficiency Partnerships Work in Other States: Is It Time for a Partnership in Colorado?" a meeting sponsored by the Colorado PUC,Denver: April 19, 1991.

18   Information about this program was obtained from personal communication with Curtis E. Gear, University of Wisconsin Cooperative Extension Service, November 1991; and

   Power and Dollars for Your Community: Economic Development Through Community Energy Programs, A Guidebook (North Central Region Rural Development Center, 1987).

19   Nancy Hirsh, "Building a Brighter Future: State Experiences in Least-Cost Electrical Planning" (Energy Conservation Coalition and the Environmental Action Foundation, August 1990), 57.

20   Ibid.

21   Cathy Ghandehari, Denver Support Office, U.S. DOE, July 1990: personal communication.

22   Ergun I. Somerson: personal communication

23   Curtis E. Gear, Wisconsin Cooperative Extension Service, November 1991: personal communication.

24   Topf Wells, WCC, November 1991: personal communication.

25   Ibid.

SUMMARY, CONCLUSIONS, AND RECOMMENDATIONS

Specific conclusions can be drawn from the case studies in this book. Although based on a limited sample, they provide helpful suggestions for legislators, as well as their support staff, to develop energy-efficiency plans:

1. Many state programs designed to improve energy-use efficiency and reduce per-capita energy consumption by energy-conservation measures have been introduced in the past few years. Because of the financial problems that most states face, state governments hope to save money and improve their economic competitiveness by improving state energy efficiency. The pace of legislation is expected to increase.

2. Many past programs have been planned and funded from oil-overcharge funds provided by the federal government. For a fraction of the cost of the programs conducted, it would have been possible to evaluate them. Unfortunately, the structure of this funding source did not require formal program evaluation as an integral part of the conservation efforts. As a result, many programs have been planned and carried out without program evaluation—especially a quantitative benefit-cost analysis—as a part of overall planning. Consequently, although billions of dollars have been spent on arguably worthwhile programs, it is not possible to quantify the level of their success. Even in those cases where quantitative information was available and benefit-cost analysis was made, the level of uncertainty in the results is high and the level of confidence in their effectiveness is low.

3. Legislators and their staff are interested in the cost effectiveness of potential energy-conservation and energy-efficiency programs. For the benefit of future legislation, it is important that legislators have available documented costs and energy savings for potential energy-efficiency programs. Most of the time, the decision about which program to implement is based on projected estimates rather than empirically verified data. Because most programs are voluntary and participation rates uncertain, in the future it will be most important to maintain records of the total effect of energy-conservation programs and their implementation costs.

4. Many state energy programs are still experimental. It is important that experience gained from these "experiments" and demonstration projects be made available nationwide. We recommend that an information-transfer system be set up between states so that the experience gained in one state can be made available to legislators in other states preparing similar legislation. As a corollary to this information-transfer system, it is also proposed that a shared data bank be established to maintain quantitative information on energy-improvement programs in different states.

5. State energy offices, particularly in small states, are often understaffed and inadequately funded. Because of the importance of maintaining accurate records and the potential benefits that can accrue from effective energy legislation, it is proposed that state energy offices receive funding appropriate to the potentially great future savings accruable from effective and successful energy programs. We also propose a close working relationship between state legislatures and state energy offices. Appendix 10 lists state energy contacts.

6. Experience gained in collaborative programs, such as those instituted in California, Iowa, and New York, should be shared with other states and key personnel involved in these collaborative processes should be made available for consultation in other states. Funding for such a consulting service might be a part of a national information-transfer service recommended under item 4. These measures will facilitate collaborative policymaking between state legislators, utility planners, public interest groups, and others.

7. Information about the various methods used to finance energy-improvement programs should be collected and disseminated. This is especially important for programs that successfully use private financing for energy-conservation measures, such as Iowa and California. Because oil-overcharge money is diminishing and will soon disappear altogether, greater emphasis should be placed on private financing of future state energy-

conservation projects.

8. Most of the energy programs deal with efficiency in buildings, but the most important challenge for energy conservation is in liquid fuels or oil. Today, the transportation sector offers the only major source for petroleum saving. It is proposed that energy savings in transportation receive more attention in future state legislation. Federal assistance to develop alternative fuels and mass transportation will be imperative to reduce liquid-fuel consumption in the states and initiate a transition from petroleum-driven transportation technologies to a transportation industry that relies on indigenous and renewable resources. In the interim, it should be observed that bus mass transit, ride sharing, and traffic-light synchronization projects are some of the most cost-effective and energy-conserving near-term options in the transportation field.

9. Legislators need objective assistance and advice from professionals who do not have potential to gain financially in the outcome of energy legislation. It is proposed that professional societies—ASME, AIEE, and AICHE, for example—provide a pool of professionals, possibly retired engineers, to help legislators in their deliberations with energy legislation.

10. Most of the energy-efficiency and conservation programs on the state level are for passive measures. Active renewable energy technologies in IRP are restricted to electric supply-side systems such as waste-to-energy combustion, and hydroelectric power plants. Demand reduction from increased use of solar thermal systems, such as domestic hot water or industrial process heat, has heretofore received insufficient IRP attention by state legislators.

11. The six case studies completed to date provide important insight into state approaches to energy problems. However, they are a limited sample and studies of energy-efficiency legislation in additional states should be performed.

12. Although reliable data on the economics of state energy conservation programs are meager, the results of those programs that were analyzed confirmed the predictions that the cost of conserved energy is below that of energy-supply options, particularly electric power.

LIST OF TERMS AND ABBREVIATIONS

ACEEE–American Council for an Energy-Efficient Economy

AEA–Area education agencies (Iowa)

AEPs–Alternate energy producers

AEPSS–Alternative Electric Power Supply Study (Wisconsin)

AFUEs–Annual fuel utilization efficiencies

AFV–Alternative fuel vehicles

AMFA–Alternative Motor Fuels Act

aMW–Average megawatts

ANWR–Arctic National Wildlife Refuge

APPA–American Public Power Association

ASHRAE–American Society of Heating, Refrigerating and Air Conditioning Engineers

AP–Advance planning

APCC–Arizona Public Corporation Commission

ARM–Agricultural Resource Management (Wisconsin)

ATTP–Automotive Technician Training Program (New York)

Appliance efficiency standard–The minimum allowable efficiency level established for a given appliance installed or offered for sale after a specified cut-off date.

BPA–Bonneville Power Administration (Washington)

BDAC–Building Design Assistance Center (Florida)

Barrel (of oil)–A volumetric unit of measure for crude oil and petroleum products equivalent to 42 U.S. gallons.

Biomass–Energy resources derived from organic matter. These include wood, agricultural waste and other living-cell material that may be burned to produce heat.

Btu–British thermal unit. A standard measure of heat content in a substance that can be burned to provide energy, such as oil, gas or coal. One Btu equals the amount of energy required to raise the temperature of one pound of water one degree Fahrenheit. A Btu is equivalent to the burning of one match stick.

c/kWh–cents per kilowatt-hour

C&LM–Conservation and load management

CO–Carbon monoxide

CO_2–Carbon dioxide

CNG–Compressed natural gas

CAFE–Corporate Average Fuel Economy Standards. Federal standards for automobile fleet fuel efficiency. In 1987, the minimum efficiency for a fleet of cars built by a single manufacturer was 26 miles per gallon.

Cogeneration–Industrial, commercial or other manufacturing systems that use heat or energy for the dual manner; that is, for a process and for generating electricity.

Commercial sector–Non-manufacturing business establishments, including hotels, motels, restaurants, wholesale businesses, retail stores, laundries, and other service enterprises; health, social and educational institutions; and federal, state, and local governments.

Conservation voltage regulation–A set of measures and operating strategies designed to provide electricity service at the lowest practicable voltage level.

DATCP–Department af Agriculture, Trade and Consumer Protection (Wisconsin)

DER–Department of Environmental Regulation (Florida)

DILHR–Department of Industry, Labor and Human Relations (Wisconsin)

DNR–Department of Natural Resources (Iowa)

DOA–Department of Administration (Iowa)

DOE–U.S. Department of Energy

DOD–Department of Defense

DOT–U.S. Department of Transportation

DSM–Demand-side management

EASI–Energy Advisory Service to Industry (New York)

EC–Externality cost

ECMs–Energy conservation measures

EDGE–Economic Development through Greater Energy Efficiency (New York)

EEMs–Energy-efficient mortgages

EES–Energy Extension Service (federal)

EFSEC–Energy Facility Site Evaluation Council (Washington)

EIA–Energy Information Agency

EILP–Energy Investment Loan Program (New York)

EMCS–Energy-management control systems

EO–Executive Order

EPA–Environmental Protection Agency

EPRI–Electric Power Research Institute

ERAM–Electricity revenue adjustment mechanism

ESHB–Engrossed Senate-House Bill

EPI–Energy performance index

Efficiency–The percentage of the total energy content of a fuel converted into useful power or heat. The remaining energy is lost to the environment as heat.

Electric utility–A corporation or other legal entity that owns and/or operates facilities for the generation, transmission, distribution, or sale of electric energy, primarily for use by the public.

Emissions–Gases and particulates from combustion discharged into the environment.

Energy–The capacity for doing work as measured by the capability of doing work (potential energy) or the conversion of this capability to motion (kinetic energy). Energy has several forms, some of which are easily convertible and can be changed to another form useful for work. Most of the world's convertible energy comes from fossil fuels burned to produce heat that is then used as a transfer medium to mechanical or other means to accomplish tasks. Electrical energy is usually measured in kilowatt-hours, while heat energy is usually measured in British thermal units.

Energy conservation–Any step taken to reduce consumption or use of energy.

Ethanol–An alcohol produced from the fermentation of biomass, usually grain and other renewable resources, such as corn, potatoes, and sugar.

FDCA–Florida Department of Community Affairs

FDER–Florida Department of Environmental Regulation

FDOT–Florida Department of Transportation

FECA–Florida Energy Conservation Act

FEECA–Florida Energy Efficiency and Conservation Act

FECSA–Florida Environmental Conservation Standards Act

FLEET–Fleet Energy-Efficient Transportation Program (New York)

FPC–Florida Power Corporation

FPL–Florida Power and Light Company

FPSC–Florida Public Service Commission

FRG–Federal Republic of Germany

FSEC–Florida Solar Energy Center

FTC–Federal Trade Commission

FY–Fiscal year

GDP–Gross domestic product

GAO–General Accounting Office

GEO–Governor's Energy Office (Florida)

GPC–Gulf Power Company

GSP–Gross state product. The total value of goods and services produced by the state's economy before deduction of depreciation charges and other allowances for capital consumption.

HIECA–Home Insulation Energy Conservation Act (New York)

HERS–Home energy rating system

HOV–High-occupancy vehicle

HVAC–Heating, ventilating and air conditioning

Hydroelectric power–Electricity generated by an electric power plant whose turbines are driven by falling water.

IBP–Institutional Buildings Program (Washington)

ICP–Institutional Conservation Program (federal)

IOU–Investor-owned utility. A utility organized under state law as a corporation to provide electric power or gas services and earn a profit for its stockholders.

IRP–Integrated resource planning

Indigenous resources–Natural resources that are not imported; that is, those produced in the same country or region where they are consumed.

Industrial sector–Manufacturing, construction, mining, agriculture, fishing, and forestry establishments.

kW–Kilowatt. One thousand watts. A unit of electric capacity or power.

kWh–Kilowatt-hour. One thousand watt hours. A unit of electricity consumed or generated.

LAN–Local area network (Washington)

LBL–Lawrence Berkeley Laboratory

LCP–Least-cost planning

LEEP–Local Energy Engineer Program (Florida)

LIHEA–Low-Income Home Energy Assistance (Nevada)

LIHEAP–Low-Income Home Energy Assistance Program

LP–Liquid propane

Least-cost integrated planning–A resource planning process that requires utilities to consider all supply and demand options, including conservation and load management, on an equal basis, and select those with the least cost for development first. In assessing the costs of an option, both direct costs and social costs should be taken into account.

Life-cycle cost analysis–Calculated as the initial cost plus the operating, maintenance and energy costs of a facility over its economic life, reflecting anticipated increases in these costs discounted to present value at the current rate for borrowing public funds..

MCS–Model Conservation Standards (Washington)

MDTA–Metro-Dade Transit Authority (Florida)

MFRA–Modified fuel revenue accounting

mpg–Miles per gallon

MSW–Municipal solid waste

MW–Megawatt. One thousand kilowatts. A unit of electric capacity or power.

Methanol–Methyl alcohol, also called wool alcohol. A gasoline additive or substitute made from coal or natural gas or biomass.

NAECA–National Appliance Energy Conservation Act (federal)

NASEO–National Association of State Energy Officials

NCSL–National Conference of State Legislatures

NEE-NET–Nevada Energy and Environmental Education Network

NEO–Nevada Energy Office

NES–National Energy Strategy

NIMBY–Not in my backyard

NOX–Nitrogen oxide

NPC–Nevada Power Company

NPPC–Northwest Power Planning Council (Washington)

NRECA–National Rural Electric Cooperative Association

NREL–National Renewable Energy Laboratory

NYPSC–New York Public Service Commission

NYSEO–New York State Energy Office

NYSERDA–New York State Energy Research and Development Authority

OPEC–Organization of Petroleum Exporting Countries

OTA–Office of Technology Assessment

PL–Public Law

PP–Payback period

PPSL–Power plant siting law (Wisconsin)

PSC–Public service commission

PUC–Public utility commission

PURPA–Public Utility Regulatory Policies Act of 1978

Photovoltaic (solar) cell–A device that converts solar radiation directly into electrical energy.

Qualifying facility–A cogeneration or small power production facility that is not more than 50 percent owned by electric utilities and meets certain standards as defined in Section 201 of Public Utility Regulatory Policies Act of 1978. These facilities produce and sell power to electric utilities.

Quad–One quadrillion Btus. The United States used about 80 quads in 1990.

RCS–Residential Conservation Service

RD&D–Research, development and demonstration (New York)

RDF–Refuse-derived fuel

RECs–Rural electric cooperatives

RFPs–Requests for proposal

R-value–A standard measure of the insulation value of various building materials or windows.

Renewable energy–A class of energy sources, such as solar, wind, hydro, and biomass, whose supply is continuously or periodically renewed.

Residential sector–Private household establishments that consume energy primarily for space heating, domestic hot water heating, air conditioning, lighting, refrigeration, cooking, and clothes drying.

SBCC–State Building Code Council (Washington)

SBEEP–Small Business Energy Efficiency Program (New York)

SECP–State Energy Conservation Program (federal)

SERI–Solar Energy Research Institute

SIFIC–State of Iowa Facilities Improvement Corporation

SO–Sulfur oxide

SO$_2$–Sulfur dioxide

SPPCo–Sierra Pacific Power Company (Nevada)

STOP–Signal Timing Optimization Program (New York)

SWMA–Solid Waste Management Act (Florida)

Simple payback–Total cost of (energy efficiency) improvements divided by annual savings due to these improvements.

TAS–Technical assistance survey (Washington)

TDM–Transportation demand management (Washington)

TEC–Tampa Electric Company

TSMP–Transportation Systems Management Program (New York)

Telecommunications–Communicating through electronic audio/visual equipment.

Thermal–Identifies a type of electric generating station or power plant, or the capacity or capability thereof, in which the source of energy for the prime mover is heat.

Total revenue cost test–(Washington)

Transportation sector–Private and public vehicles that move people and commodities. Included are automobiles, trucks, buses, motorcycles, railroads and railways, aircraft, ships, barges and natural gas pipelines.

UNLV–University of Nevada, Las Vegas

UTC–Utilities and Transportation Commission (Washington)

UW–University of Wisconsin

VMT–Vehicle miles traveled

WAP–Weatherization Assistance Program (Nevada)

WAPA–Western Area Power Administration

WCC–Wisconsin Conservation Corps

WCDSR–Wisconsin Center for Demand-Side Research

WEB–Wisconsin Energy Bureau

WSEC–Washington State Energy Code

WSEO–Washington State Energy Office

W–Watt. An electric unit of power equivalent to one ampere flowing under a potential of one volt.

Weatherization–The tightening and insulating of a structure to lower its energy consumption.

ZEV–Zero-emission vehicles

Appendix 1
1989 Gasoline Consumption Per State & Per Capita In Each State

State	Total Gasoline Use (thousands of gallons)	Population Per State	Per Capita Gas Use (gallons)
Alabama	2,119,236	4,062,608	521.64
Alaska	235,365	551,947	426.43
Arizona	1,744,552	3,677,985	474.32
Arkansas	1,260,344	2,362,239	533.54
California	13,332,687	29,839,250	446.82
Colorado	1,520,095	3,307,912	459.53
Connecticut	1,380,346	3,295,669	418.84
Delaware	349,087	668,696	522.04
D.C.	179,437	609,909	294.20
Florida	6,112,751	13,003,362	470.09
Georgia	3,578,425	6,508,419	549.81
Hawaii	384,743	1,115,274	344.98
Idaho	484,849	1,011,986	479.11
Illinois	4,943,724	11,466,682	431.14
Indiana	2,646,931	5,564,228	475.70
Iowa	1,395,852	2,787,424	500.77
Kansas	1,281,262	2,485,600	515.47
Kentucky	1,857,461	3,698,969	502.16
Louisiana	2,007,477	4,238,216	473.66
Maine	608,991	1,233,223	493.82
Maryland	2,101,878	4,798,622	438.02
Massachusetts	2,494,831	6,029,051	413.80
Michigan	4,331,378	9,328,784	464.30
Minnesota	2,081,338	4,387,029	474.43
Mississippi	1,245,703	2,586,443	481.63
Missouri	2,729,701	5,137,804	531.30
Montana	444,020	803,655	552.50
Nebraska	790,760	1,584,617	499.02
Nevada	636,935	1,206,152	519.78
New Hampshire	526,087	1,113,915	472.29
New Jersey	3,482,794	7,748,634	449.47
New Mexico	810,933	1,521,779	532.88
New York	5,706,382	18,044,505	316.24
North Carolina	3,331,646	6,657,630	500.43
North Dakota	359,970	641,364	561.26
Ohio	4,974,535	10,887,325	456.91
Oklahoma	1,667,680	3,157,604	528.15
Oregon	1,366,267	2,853,733	478.76
Pennsylvania	4,659,905	11,924,710	390.78
Rhode Island	380,843	1,005,984	378.58

Appendix 1 (continued)
1989 Gasoline Consumption Per State & Per Capita In Each State

State	Total Gasoline Use (thousands of gallons)	Population Per State	Per Capita Gas Use (gallons)
South Carolina	1,826,586	3,505,707	521.03
South Dakota	393,202	699,999	561.72
Tennessee	2,572,738	4,896,641	525.41
Texas	8,725,904	17,059,805	511.49
Utah	743,572	1,727,784	430.36
Vermont	280,654	564,964	496.76
Virginia	3,033,215	6,216,568	487.92
Washington	2,306,241	4,887,941	471.82
West Virginia	833,899	1,801,625	462.86
Wisconsin	2,103,872	4,906,745	428.77
Wyoming	324,463	455,975	711.58
Total	114,681,547	249,632,692	

Source: U.S. Department of Transportation, Federal Highway Administration, *Highway Statistics* 1989, 17 (1989).
U.S. Department of Commerce, Bureau of the Census, *Commerce News*, 2 (Dec. 26, 1990).

Appendix 2
1989 Vehicle Miles Traveled Per State & Per Capita In Each State

State	1989 VMT (millions of miles)	Population Per State	VMT Per Capita
Alabama	40,765	4,062,608	10,034.19
Alaska	3,887	551,947	7,042.34
Arizona	34,816	3,677,985	9,466.05
Arkansas	20,414	2,362,239	8,641.80
California	251,482	29,829,250	8,427.89
Colorado	25,577	3,307,912	8,336.68
Connecticut	26,183	3,295,669	7,944.67
Delaware	6,446	668,696	9,639.66
D.C.	3,414	609,909	5,597.56
Florida	108,877	13,003,362	8,372.99
Georgia	75,705	6,508,419	11,631.86
Hawaii	7,750	1,115,274	6,948.97
Idaho	8,422	1,011,986	8,322.25
Illinois	81,297	11,466,682	7,089.85
Indiana	56,192	5,564,228	10,098.80
Iowa	22,571	2,787,424	8,097.44
Kansas	21,913	2,485,600	8,815.98
Kentucky	32,165	3,698,969	8,695.67
Louisiana	37,914	4,238,216	8,945.75
Maine	11,739	1,233,223	9,518.96
Maryland	38,922	4,798,622	8,111.08
Massachusetts	46,214	6,029,051	7,665.22
Michigan	79,890	9,328,784	8,563.82
Minnesota	37,393	4,387,029	8,523.54
Mississippi	22,895	2,586,443	8,851.93
Missouri	48,087	5,137,804	9,359.45
Montana	8,250	803,655	10,265.60
Nebraska	13,781	1,584,617	8,696.74
Nevada	9,408	1,206,152	7,800.01
New Hampshire	9,819	1,113,915	8,814.86
New Jersey	59,898	7,748,634	7,730.14
New Mexico	15,839	1,521,779	10,408.21
New York	106,059	18,044,505	5,877.63
North Carolina	60,877	6,657,630	9,143.94
North Dakota	5,849	641,364	9,119.63
Ohio	84,418	10,887,325	7,753.79
Oklahoma	32,836	3,157,604	10,399.02
Oregon	25,820	2,853,733	9,047.80
Pennsylvania	83,855	11,924,710	7,032.04
Rhode Island	6,740	1,005,984	6,699.91
South Carolina	32,780	3,505,707	9,350.47

302

Appendix 2 (continued)
1989 Vehicle Miles Traveled Per State & Per Capita In Each State

State	1989 VMT (millions of miles)	Population Per State	VMT Per Capita
South Dakota	6,704	699,999	9,577.16
Tennessee	45,639	4,896,641	9,320.47
Texas	159,512	17,059,805	9,350.17
Utah	13,915	1,727,784	8,053.67
Vermont	5,765	564,964	10,204.19
Virginia	59,337	6,216,568	9,544.98
Washington	43,233	4,887,941	8,844.83
West Virginia	14,940	1,801,625	8,292.51
Wisconsin	43,086	4,906,745	8,780.97
Wyoming	5,750	455,975	12,610.34
Total	2,107,040	249,632,692	

Source: U.S. Department of Transportation, Federal Highway Administration, *Highway Statistics 1989*, 182 (1989); U.S. Department of Commerce, Bureau of the Census, *Commerce News*, 2 (Dec. 26, 1990).

Appendix 3
Lawrence Berkeley Laboratory
Report on Residential and Commercial Conservation

This report was written by James E. McMahon, Lawrence Berkeley Laboratory, on March 2, 1990, under the title "Supply Curves of Conserved Energy: Residential and Commercial Sectors." It was published as attachment E in *Preliminary Technology Cost Estimates of Measures Available to Reduce U.S. Greenhouse Gas Emissions by 2010*, a report submitted to the U.S. Environmental Protection Agency, August 1990, by ICF Incorporated, 9300 Lee Highway, Fairfax, Va. 22031. It is not included in the copyright of this book.

Introduction

1. Principles of Supply Curves of Conserved Energy
2. Methodology
   2.1 Residential Sector
   2.2 Commercial Sector
   2.3 Administrative Costs
3. Technology Data Base
4. Current Market
5. Projections
6. Results: Costs of Conserved Energy

Introduction

Section 1 describes supply curves of conserved energy.

Section 2 describes the methodology used in this study to estimate costs of conserved energy for the residential and commercial sectors, including a brief discussion of the administrative costs which are added to the technology costs.

Section 3 describes the technology data base.

Section 4 discusses the current market, including a list of appliances currently bearing energy efficiency labels. Market barriers to adoption of energy efficient technologies are listed.

Section 5 describes the base case projections, from which additional conservation costs and savings are calculated.

Section 6 presents the results, namely costs of conserved energy and energy savings, for the years 2000 and 2010 for the residential and commercial sectors.

1. Principles of Supply Curves of Conserved Energy

The supply curve of conserved energy is a useful tool in least-cost utility planning. It graphically portrays the technical potential for energy conservation in a way that is easy to grasp and assess (**figure 1**). Just as important, however, is the consistent assumptions framework underlying the supply curve. The consistency of assumptions regarding costs, energy savings, lifetimes, and other key factors simplifies the comparison of individual conservation measures. More generally, the methodology permits comparison with new energy supplies. A table—often a spreadsheet— provides the backup information for the supply curve. The assumptions behind the supply curve of conserved energy are presented below. Many of the complications of the approach are not described in order to present the overall approach; more detailed discussions are listed in the bibliography. The same approach can be used with only minor modification for assessing measures to reduce water, CO_2, and other resources.

Figure 1
Energy Supplied Through Conservation

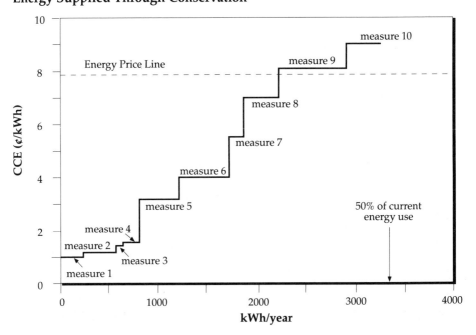

Figure 1. A hypothetical supply curve of conserved electricity. Each step represents a conservation measure. The width of the measure represents the energy savings (in kWh/year), and the height indicates the cost of conserved energy (CCE). A measure is cost-effective if its CCE is less than the price of the energy it displaces. Thus, all measures below the "energy price line" are cost-effective. This is a "micro" supply curve, that is, for one water heater, a refrigerator, or the space heating heating system in a house. A table usually accompanies the supply curve, which includes a longer description of the measure, its cost, energy savings, and CCE.

The Cost of Conserved Energy

The cost of conserved energy (CCE) is the measure of economic value of a conservation measure in a supply curve of conserved energy and it represents the vertical axis of the conservation supply curve. The CCE is an investment statistic. The CCE is similar to return on investment, payback time, and internal rate of return except that it relies on slightly different information.

The cost of conserved energy is defined by the formula:

$$\text{CCE} = \frac{\text{Investment}}{\text{Energy Savings}} \times \frac{d}{1-(1+d)^{-n}}$$

The formula consists of two parts. The first part is simply the ratio of the conservation measure's cost over the savings (usually expressed per year). The second part is an annuity factor. This factor converts the measure's cost into an annual payment, based on the lifetime, n, and discount rate, d. The dimensions of the CCE depend on the units used in the CCE calculation. For example, if the cost is expressed in dollars, the energy savings in kilowatt-hours per year, the interest rate per year, and the lifetime in years, then the resulting CCE will have the dimensions of \$/kWh. The intuitive meaning of the CCE is the cost to save a unit of energy. The cost of conserving other resources can also be calculated by substituting that resource, such as tons of avoided CO_2 or gallons of saved water, in place of electricity savings.

For each measure in the supply curve, one must specify the cost, energy savings and lifetime (The discount rate is assumed to be the same for all measures). This requirement already imposes a degree of consistency among conservation measures and permits one to compare CCEs of measures. The most attractive measures are those with lower CCEs. Ranking measures by increasing CCE establishes an order of economic attractiveness, the most attractive first. This procedure establishes the sequence of conservation measures on the supply curve.

The CCE alone gives no information about cost-effectiveness. To decide if a conservation measure is cost-effective, the CCE must be compared to the price of the energy that is avoided. Note that the CCE and energy price will have the same dimensions (if the inputs are correctly chosen). If the cost of conserving a unit of energy is less than that of the displaced energy, then a measure is cost-effective. For example, a refrigerator efficiency improvement might have a CCE of \$0.02/kWh. (The cost of the measure was dollars and the energy savings in kWh, hence the CCE is \$/kWh.) To decide if this measure is cost-effective, compare to the price of electricity the avoided electricity use, say, \$0.10/kWh. On a supply curve of conserved energy, an "energy price line" can be drawn across until it intersects with measures having CCEs higher than the price (**figure 1**). All measures below the energy price line are cost-effective; those above the line are not.

Micro and Macro Supply Curves

The simplest supply curve of conserved energy consists of a collection of unrelated conservation measures. In this case, the measures can be stacked, in order of increasing CCE, on the supply curve. Each measure is represented by a step whose width is the energy saved (per year) and whose height is the CCE. This is sometimes called a "micro" supply curve of conserved energy because each step represents savings from one application of that measure; that is, one furnace, one car, or one light bulb. In contrast, a step in a "macro" supply curve may represent the average savings and CCE for the measure in thousands of different furnaces, cars, or light bulbs.

Most micro supply curves are more complicated than the one first described because the measures are connected. For example, a typical supply curve might represent the conservation measures that could be applied to a residential water heating system. In this case, the energy savings are no longer independent because the savings from one measure will often depend on the measures that have already been implemented. The energy savings attributed to an improvement in the water heater's efficiency will depend on the amount of hot water demanded which, in turn, will depend on the measures that have already been implemented (such as a low-flow showerhead). Put another way, the sum of savings of each measure implemented alone will be greater than the two implemented together. If the interdependence of the measures is not taken into account, then it is possible to "double-count" the energy savings. Under extreme circumstance, it is possible to demonstrate that more energy can be saved than was actually used in the first place (an embarrassing result). Sometimes the interdependence of measures can save more energy than the measures applied singly. Improving the efficiency of lighting in commercial buildings, for example, may save lighting and air conditioning energy.

A properly constructed supply curve of conserved energy will avoid double-counting errors by the following procedure. The CCE is calculated for all of the measures. The cheapest (lowest CCE) measure is selected and "implemented;" that is, the energy savings from the first measure are subtracted from the initial energy use. The new energy use is used to recalculate the CCEs of the remaining measures. (In general, the CCEs will rise.) The measure with the lowest CCE is selected and implemented. The energy use is recalculated, along with the CCEs for this lower energy use. The procedure is repeated until all the measures have been ranked. This procedure has several implications about energy savings of measures when taken out of order; this is discussed in Meier (1982).

A key assumption in the supply curve of conserved energy is the initial energy use (or baseline). Since the supply curve shows reductions in energy use from the baseline, it is crucial to carefully define the baseline. It is very difficult to save energy that was not used in the first place. Determining the initial energy use, and the energy-related characteristics, is often the most difficult part of constructing a conservation supply curve. Measured data is obviously superior but more expensive to obtain.

Macro Supply Curves

A "macro" supply curve of conserved energy refers to a combination of micro supply curves. Thus, one might construct a macro supply curve for all the residential electric water heaters in Massachusetts or central air conditioners (actually the end use) in Texas. It is impossible to evaluate and aggregate the savings from every unit in a large collection. Instead, prototypes are created that represent the average micro situation. Then, with the assistance of engineering calculations, simulations, or even measured data, average savings are estimated for a collection of conservation measures. Those savings for an individual unit are then multiplied by the number of units (water heaters in Massachusetts, air conditioners in Texas, for example) to obtain the macro energy savings (**figure 2**).

A macro supply curve of conserved energy offers two insights into the role of energy efficiency. Is energy conservation worth pursuing? One can easily recognize the overall significance of energy efficiency to present energy supplies by comparing the potential savings to current use. Which measures save the most and which are the cheapest? The relative importance of conservation measures are shown both with respect to their potential contribution of energy savings (the width of each measure's step) and the relative costs (the height of each measure's step). The consistent accounting framework behind the supply curve ensures that the measures are indeed comparable.

Since it is impossible to measure each unit in a macro supply curve, statistical methods are needed to accurately characterize each measure and the eligible population. Such information is generally derived from surveys, the census, and production data rather than simple engineering analyses. Further complications are introduced if the stock of energy-using equipment is expected to change over time due to retirements, standards, or population increase. All of these considerations introduce new kinds of uncertainties. The accurate characterization of the baseline—often the baseline over time—becomes increasingly important.

Supply curves of conserved energy portray the technical potential for saving energy but a more useful estimate is the amount of conservation that can be reasonably achieved. It is possible to superimpose estimates of the

Figure 2
Cumulative Energy Supplied

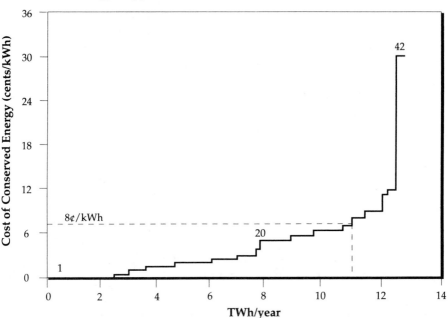

Figure 2. A "macro" supply curve of conserved electricity for California's residential sector. Taken from Meier, Wright, and Rosenfeld (1983).

fraction of the stock that could be changed and the costs of the programs to reach them. Similar procedures can be used to gradually phase in conservation over several years. Graphically, these actions result in a supply curve that is horizontally squashed and above the original technical potentials curve. Again, the results are easy to visualize and interpret.

Supply Curves—A Tool With Limitations

The supply curve of conserved energy is popular because it is a simple way to assess the potential for energy conservation and the relative impacts of conservation measures. At the same time, the methodology has distinct limitations. Supply curves are often inappropriate when more than one fuel type is involved. In electricity studies, the supply curve methodology cannot easily combine the benefits of savings in energy and demand. The supply curve approach measures savings from a baseline. If the baseline does not exist (or is poorly understood) then calculating savings from it is likely to generate large errors. For these (and other) reasons, supply curves of conserved energy are probably best applied at middle range of detail. For crude analyses, the curves do not provide any more insights than, say, tables. For detailed studies, the limitations of the supply curves prevent an integrated analysis.

Semantics

The "supply curve of conserved energy" is a misnomer, but has nevertheless become the accepted term. It is not a supply curve in the traditional sense because it is not a proven response; the market does not offer the conserved energy as the price of supplied energy varies. It is to some extent a price schedule for a resource. Others prefer to call it a production function.

2. Methodology

This section describes the method used to produce draft conservation supply curves for the residential and commercial sectors, sent to EPA on January 23-24. *All costs of conserved energy, unless labeled as "Technology Only," include a 20 percent markup to account for implementation costs. This applies to all figures.*

2.1 Residential Sector

The technology data base for the residential sector is the set of designs identified for each appliance as part of the analysis of federal appliance energy conservation standards. This data base is used by the LBL Residential Energy Model (REM), which produces projections to the year 2030 of unit energy consumption, annual sales of appliances, and total energy consumption for the U.S. The base case projection from the LBL REM includes appliance standards already on the books through 1988, affecting new appliances beginning in 1990. (The most recent updates, which apply to refrigerators and freezers beginning in 1993, are not included.) The base case from which conservation potential was assessed assumed efficiencies of new appliances frozen at the 1990 level. (A base case in which efficiencies changed as a function of projected changes in energy prices differs only slightly from the frozen efficiency case, since projected electricity prices change little in the forecast period.)

The data for the conservation supply curves are produced by spreadsheets containing the technology data base, in one case, and building shell measures, in the other curve. These spreadsheets contain the same set of appliance designs or shell measures as used in the LBL REM. They also take in projections of appliances and number of houses in stock for a particular year from LBL REM. Based on those projected quantities, the possible energy savings and costs are produced as a table. The measures are then ordered by ascending cost of conserved energy. A graph is plotted from this table.

The graph shows, on the vertical axis, cost of conserved energy, expressed as 1988 dollars per million Btu (primary energy). For electricity, 11,500 Btu equals one kWh. The horizontal axis is energy savings, in units of trillion Btu (primary). The energy savings are for measures implemented by the year indicated, accumulated over the options possible. In other words, the cost of conserved energy is obtained from the technology (or building shell) data base for each measure beyond the base case. The energy savings for each measure are obtained as the difference between a base case projection of average unit energy consumption (UEC) in stock and the UEC of a more efficient option in the data base, times the number of appliances (or buildings) projected to be stock in the year (2000 or 2010). Cumulative energy savings for the year are the sum over options.

Electricity and gas savings are presented on separate graphs for appliances. For shell measures, separate tables and graphs are presented for electrically heated homes, and for gas-heated homes. For electrically heated homes, energy savings include both heating and cooling energy, where energy savings are calculated as weighted sum over technologies. For example, for cooling, the technologies are room air conditioners, central air conditioners, and heat pumps. Heating technologies are resistance and heat pumps. For shell measures applied to gas-heated dwellings, the cost of conserved energy attributes all the costs to gas savings. Displayed in the tables, but not in the figures, are the electricity savings in gas-heated homes, from effects of the shell measures on air conditioning.

Cost of conserved energy for retrofit measures (to existing building shells) has been estimated by assuming that the measures are applied to the entire building stock. No consideration is given to constraints imposed by the normal turnover rate of appliances or buildings, or by the rate at which retrofits have occurred in the past.

2.2 Commercial Sector

The technology data base for the commercial sector is the set of designs identified for each end use and building type in the ACEEE report, *The Potential for Electricity Conservation in New York State*, (P.M. Miller, J.H. Eto, and H.S. Geller, 1989). Only electricity is considered. This data is used by the PNL Commercial Energy Model (kept by D. Belzer) which produces projections to the year 2010 of end-use energy utilization intensity (EUI), annual floorspace by building type, and total energy consumption for the United States. The base case from which conservation potential was assessed is the same as the "Where We Are Headed" scenario in the National Energy Strategy paper (R. Carlsmith , et al., 1989. "Energy Efficiency: How Far Can We Go?" ORNL/...). The base case in which efficiencies changed as function of projected changes in energy prices differs only slightly from a frozen efficiency case, since projected electricity prices change little in the forecast period.

The data for the conservation supply curves are produced by a spreadsheet containing the technology and building shell data base. This spreadsheet contains the percent savings, by end use and building type, expected for each measure, according to the New York study. The difference between those savings and the base case savings are reported in the conservation supply curve. The spreadsheet takes in projections of floorspace and EUI by building type from the base case forecast. Based on those projected quantities, the possible energy savings and costs are produced as a table. The costs of each measure are averaged over the building types, weighted by energy savings, then placed in ascending order of cost of conserved energy. A graph is plotted from this table.

The graph shows, on the vertical axis, cost of conserved energy, expressed as 1988 dollars per million Btu (primary energy). For electricity, 11,500 Btu equals one kWh. Only electricity savings are on the graph. The horizontal axis is energy savings, in units of trillion Btu (primary). The energy savings are for measures implemented by the year indicated, accumulated over the options possible. In other words, the cost of conserved energy is obtained from the technology (or building shell) data base for each measure beyond the base case. The energy savings for each measure are obtained as the difference between a base case projection of EUI, and EUI of a more efficient option in the data base, times the projected floorspace for the building type in this year. Cumulative energy savings for the year are the sum over options.

No attempt has been made to eliminate double-counting. The correction factor would be the difference between the sum of energy savings due to upgrading space conditioning equipment and shell measures. We expect this factor to be small.

Cost of conserved energy for retrofit measures (to existing building shells) has been estimated by assuming that the measures are applied to the entire building stock. No consideration is given to constraints imposed by the normal turnover rate of equipment or buildings, or by the rate at which retrofits have occurred in the past.

2.3 Administrative Cost

In addition to technology costs, implementation of conservation measures involves additional costs. The administrative costs of implementing any single conservation measure depend upon the approach taken. For example, an advertising campaign will incur different costs than an audit program, even if both are intended to increase the purchase of ceiling insulation.

No attempt was made to characterize the program costs appropriate to each conservation measure here. However, it was felt inappropriate to completely ignore program costs. Therefore, as an approximation to the average administration costs of all conservation measures, all costs of conserved energy were increased by 20 percent (over the technology cost) to account for administrative cost.

3. Technology Data Base

The residential technology data base is comprised of the retail price, energy efficiency, and unit energy consumption of alternative designs of residential appliances (and building shells) used by the LBL REM for analysis of DOE appliance performance standards.[1] The technology data base contains data for the following appliances:

Electric appliances: central air conditioners, heat pumps, room air conditioners, incandescent lighting, water heaters, refrigerators, freezers, dishwashers, clothes washers, clothes dryers, televisions, and pool controls.

Gas appliances: furnaces, water heaters, and clothes dryers.

In the lexicon of DOE appliance standards, the technology data base for each end use may be comprised of several product classes, each containing many design options. For example, a product is a refrigerator. There are seven product classes for refrigerators and refrigerators/freezers, differing by the type of defrost system (manual, partial, or automatic), the placement of the doors (freezers on top, bottom, or side), and the presence, or absence, of through-the-door features.

• Manual defrost refrigerator,
• Partial automatic defrost refrigerator/freezer,
• Top-mount auto-defrost refrigerator/freezer without through-the-door features,
• Top-mount auto-defrost refrigerator/freezer with through-the-door features,
• Side-by-side auto-defrost refrigerator/freezer without through-the-door features,
• Side-by-side auto-defrost refrigerator/freezer with through-the-door features,
• Bottom-mount auto-defrost refrigerator/freezer.

A design option is a specific combination of components, differing in energy efficiency. For each product class of refrigerators, there are 7 of 12 design options applicable. Some of the design options considered for refrigerators and freezers include:

- Enhanced evaporator heat transfer,
- More efficient compressor,
- Foam (to replace fiberglass) in refrigerator door,
- Thicker insulation in doors,
- More efficient fan,
- Evacuated panel insulation (replaces foam),
- Two-compressor system,
- Adaptive defrost.[2]

The design options for other products are listed in the table of results for appliances, ranked by cost of conserved energy.

Each design option stands as a conservation measure. If refrigerators in place in 2000 were replaced with units characterized by lower unit energy consumption, the energy savings can be characterized as the difference in unit energy consumption (between the average unit in stock and the design option proposed as a conservation measure) times the number of units in stock. The CCE associated with this conservation measure is the CCE design option. Additional energy savings are calculated as the difference in UEC between design option with the next higher CCE and the previous design option, again times the number of units in stock.

Building shell measures. In addition to replacing equipment, measures which conserve energy by improving the thermal performance of building shells are considered. These measures include increasing floor insulation (to R-19), decreasing infiltration (to 0.4 air changes per hour [ACH]), increasing ceiling insulation (to R-31, R-42, or R-50), increasing wall insulation (to R-11), and increasing glazing (to triple-pane windows).

All building shell measures were applied to the average stock house in the year indicated (either 2000 or 2010). In other words, all building shell measures were handled as if they were retrofits, rather than identifying measures that could be applied to new homes in the intervening years. In addition, Washington, D.C., weather was assumed typical of national average weather.

Double-counting. When compiling a list of energy conservation measures which mutually impact a particular end-use—such as space heating—the savings attributed to any conservation measure must be corrected for any savings already attributed to other measures previously implemented. For example, if both a more efficient furnace *and* increased ceiling insulation are potential conservation measures, they must be considered in order, and the energy consumption adjusted to account for the effects of the first measure taken before calculating the energy savings from the second measure. Failure to follow this procedure will lead to overestimating the total energy savings (called "double-counting").

No corrections were made for double-counting in the commercial sector. In the results reported here for the residential sector, the equipment measures were generally observed to have lower CCE than the building shell measures. Consequently, the energy consumption was corrected for equipment conservation measures before calculating the CCE for the building shell measures. In other words, the more efficient furnaces, air conditioners, and heat pumps were assumed when calculating the additional savings possible from building shell measures.

Commercial Sector. For the commercial sector, the set of potential energy conservation measures was taken from a recent study for the state of New York.[3]

4. Current Market

The current market for residential appliances includes a range of efficiencies available for purchase. Some appliances carry labels on each model showing the efficiency or average operating cost, in order to provide information to purchasers. Those appliances requiring such labels are shown in **table 1**.

In addition, **table 1** indicates those appliances for which national energy performance or efficiency standards have been promulgated. For some appliances, there are both energy performance standards and labeling requirements.

For most appliances, trade association public directories list characteristics, including the energy efficiency of each model. From these directories, one can determine the range of efficiencies currently available. **Table 2** shows the most efficient models available, for typical classes and sizes.

5. Projections

The supply curve of conserved energy for the residential sector was calculated from a frozen efficiency base case. Efficiencies of new units sold after 1990 were constant at the 1990 level, taking into account existing appliance standards regulations. (The energy performance standard for 1993 refrigerators and freezers was not included.) For the commercial sector, a base case from the PNL Commercial Energy Model was used.

Business-as-usual efficiency improvements. In the commercial sector, the business-as-usual penetrations of more efficient technologies are contained in the base case. In the residential sector, the projection assumes that little improvement beyond the mandatory energy performance standards will occur by 2010. The view is supported by comparing the frozen efficiency case

Table 1
Regulated Residential Appliances

Appliance	Energy Standard	Label
BOTH ENERGY STANDARDS AND LABELS		
1. Refrigerators, freezers	X	X
2. Room air conditioners	X	X
3. Central air conditioners/heat pumps	X	X
4. Water heaters	X	X
5. Furnaces	X	X
6. Dishwashers	X	X
7. Clothes washers	X	X
ENERGY STANDARDS; NO LABELS		
8. Clothes dryers	X	–
9. Direct heating equipment	X	–
10. Kitchen ranges and ovens	X	–
11. Fluorescent light ballasts	X	–
NO ENERGY STANDARDS; NO LABELS		
12. Pool heaters	–	–
13. Television sets	–	–
14. Humidifiers and dehumidifiers	–	–

Table 2
Most Energy-Efficient Models

Refrigerator	top freezer, 18.6 cu ft	840	kWh/yr
Freezer	chest, manual defrost, 20.7 cu ft	528	kWh/yr
Dishwasher		574	kWh/yr
Clotheswasher	front loading	451	kWh/yr
Clotheswasher	top loading	651	kWh/yr
Water heater	gas, 38 gallon	.65	Energy Factor
Water heater	electric, 50 gallon	.96	Energy Factor
Water heater	heat pump, 52 gallon	3.5	Energy Factor
Room air conditioner	10,100 Btu/hr	12.0	EER
Central air conditioner	3 tons	15.0	SEER
Central heat pump	3 tons	11.3	SEER
		8.50	HSPF
Gas furnace	77,000 Btu/hr	96.0	AFUE (%)

Source: American Council for an Energy-Efficient Economy, *The Most Energy-Efficient Appliances*, 1988 edition, Washington, D.C.

to a business-as-usual projection by the LBL Residential Energy Model. Given current EIA projections of residential electricity prices, which show little increase in real price over time, the projection shows little increase in energy efficiency.

The LBL REM incorporates coefficiencies derived empirically from observed market behavior over the past 15–20 years. Particularly in the area of energy-efficiently choice, there are a number of formidable market barriers. These are discussed in more detail elsewhere,[4] but include:

1. Indirect or forced purchase decisions (builders or landlords selecting appliances for which they do not pay the operating costs, or emergency replacements of malfunctioning equipment);

2. Lack of clear information about costs and benefits of energy efficiency improvements;

3. Purchasers may lack sufficient capital to purchase a more energy-efficient product;

4. Purchasers may have a threshold below which savings may not be significant or worth the additional effort to obtain;

5. Highest efficiency designs may not be universally available, or may be bundled with other features;

6. Manufacturers' decisions to improve energy efficiency are often secondary to other design changes. and may take years to implement;

7. Marketing strategies by manufacturers or retailers may intentionally lead to sales of less efficient equipment.

Variations in energy prices or usage rates. Across the United States, there are wide variations in the prices of energy from locality to locality. According to a recent projection,[5] at the state level residential electricity prices range from 4.5 cents per kilowatt-hour in Idaho to 11.7 cents/kWh in New York. In addition, differences in household size (persons per household), occupancy patterns (Is the house occupied during the day?), and usage patterns (such as, thermostat settings) lead to wide variations in household energy consumption patterns. The analysis performed to date assumes average energy prices, average usage behavior, and average weather. A more detailed analysis is called for determine the differences among regions. (LBL is engaging in such a study for 10 federal regions in FY 1990.)

In addition, sensitivity analyses are needed to determine the range of cost of conserved energy for each conservation measure, as a function of differences in usage rates. These studies are most important where variation is the greatest, and may be less important, although not negligible, for some end-uses where usage behavior is less of a determining factor (perhaps refrigerators).

6. Results: Costs of Conserved Energy

All costs of conserved energy, unless labeled as Technology Only, included a 20 percent markup to account for implementation costs. This applies to all figures. The costs of conserved energy for each of the conservation measures considered are presented in tables and figures. **Figure 3** and **table 3** show the cost of conserved energy for the commercial sector in 2000. The cost of each measure is averaged over all the building types.) **Figure 4** and **table 4** show the cost of conserved energy for the commercial sector in 2010.

Figure 5 and **table 5** show the supply curve of conserved energy for the residential sector in 2000 for electric appliances. **Figure 6** and **table 6** show the supply curve of conserved energy for residential building shell measures in 2000 for electrically heated and cooled houses. **Figure 7** and **table 7** show the supply curve of conserved energy for the residential sector in 3000 for gas appliances. **Figure 8** and **table 8** show the supply curve of conserved energy for residential building shell measures in 2000 for gas heated houses.

Figure 9 and **table 9** show the supply curve of conserved energy for the residential sector in 2010 for electric appliances. **Figure 10** and **table 10** show the supply curve of conserved energy for the residential building shell measures in 2010 for electrically heated and cooled houses. **Figure 11** and **table 11** show the supply curve of conserved energy for residential sector in 2010 for gas appliances. **Figure 12** and **table 12** show the supply curve of conserved energy for residential building shell measures in 2010 for gas heated houses.

Figure 3
Conservation Supply Curve
Commercial Energy 2000

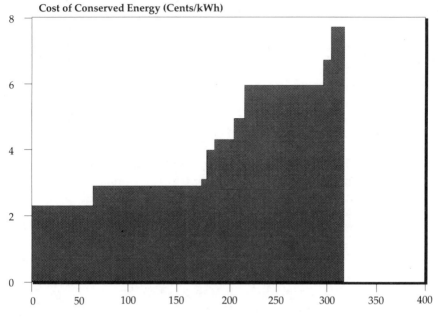

Cost of Conserved Energy (Cents/kWh)

Cumulative Energy Saved (TWh)

Table 3
Energy Supply Curve for Commercial Buildings in 2000

Energy Saved (TWh)	Technology Only CCE (cent/kWh)	Technology & Program CCE (cent/kWh)	Energy Measure
22.3	2.05	2.46	high efficiency ballasts
38.5	2.10	2.52	reflectors for fluorescent lamps
.4	2.40	2.88	re-set supply air temperature
8.1	2.47	2.96	economizer
102.9	2.51	3.01	adjustable speed drives for fan motors
4.1	2.74	3.29	energy saving fluorescent lamp
6.6	3.40	4.08	vav conversion
24.7	3.78	4.54	occupancy sensors
6.4	4.39	5.27	adjustable speed drives for pumps
.7	4.60	5.52	high efficiency fan motors
3.6	4.72	5.66	window films
.9	4.80	5.76	refrigerator case covers
74	5.14	6.17	daylighting controls
5.5	5.80	6.96	re-size chillers
.1	6.20	7.44	high efficiency pump motors
12.9	6.46	7.75	very high efficiency lamps and ballasts

| 311.7 | TOTAL |

Figure 4
Conservation Supply Curve
Commercial Energy 2010

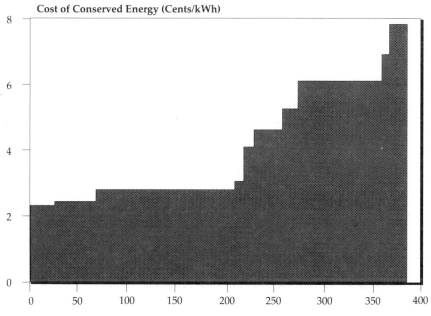

Cost of Conserved Energy (Cents/kWh)

Cumulative Energy Saved (TWh)

Table 4
Energy Supply Curve for Commercial Buildings in 2010

Energy Saved (TWh)	Technology Only CCE (cent/kWh)	Technology & Program CCE (cent/kWh)	Energy Measure
27.8	2.03	2.44	high efficiency ballasts
46.1	2.10	2.52	reflectors for fluorescent lamps
.5	2.40	2.88	re-set supply air temperature
9.8	2.47	2.96	economizer
126.1	2.51	3.01	adjustable speed drives for fan motors
5.1	2.74	3.29	energy saving fluorescent lamp
8.3	3.40	4.08	vav conversion
30.7	3.78	4.54	occupancy sensors
7.7	4.40	5.28	adjustable speed drives for pumps
.9	4.60	5.52	high efficiency fan motors
4.5	4.72	5.66	window films
.8	4.80	5.76	refrigerator case covers
93.7	5.14	6.17	daylighting controls
6.9	5.80	6.96	re-size chillers
.2	6.20	7.44	high efficiency pump motors
15.1	6.54	7.85	very high efficiency lamps and ballasts

384.2   TOTAL

Figure 5
Conservation Supply Curve
Residential Energy 2000

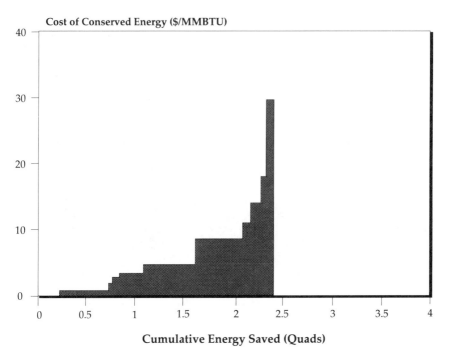

Cost of Conserved Energy ($/MMBTU)

Cumulative Energy Saved (Quads)

Electric Appliances Only

Table 5
Residential Conservation Supply Curves 2000 – Electric Appliances

Appliance	Level	Design	UEC (kWh/y)	SEER	Consumer Price	Technology Only CCE ($/MMBtu)	Technology & Program CCE ($/MMBtu)	Per Unit Energy Savings (MMBtu/y)	Total Energy Savings (TBtu/y)	Cum. Energy Savings (TBtu/y)
Central Air	0	Baseline	2884.9	9.52	1607.00	NA	NA	NA	NA	NA
Dishwasher	0	Baseline	839.0		306.58	NA	NA	NA	NA	NA
Heat Pump	0	Baseline	9867.8	9.46	1770.98	NA	NA	NA	NA	NA
Lighting	0	Baseline	1000.0		4.72	NA	NA	NA	NA	NA
Electric Hot Water	0	Baseline	4157.1	89.1	280.46	NA	NA	NA	NA	NA
Electric Clothes Dryer	0	Baseline	988.0	158.00	311.60	NA	NA	NA	NA	NA
Chest Freezer	-1	Stock Unit	578.7	149.00	347.02	NA	NA	NA	NA	NA
Pool Controls	0	Baseline	2500.0		0.00	NA	NA	NA	NA	NA
Clothes Washer	0	Baseline	926.1	190.00	380.04	NA	NA	NA	NA	NA
Room Air	0	Baseline	569.7	8.72	382.00	NA	NA	NA	NA	NA
Refrigerator	-1	Stock Unit	1052.0	224.00	521.70	NA	NA	NA	NA	NA
Upright Freezer	-1	Stock Unit	800.1	159.00	370.31	NA	NA	NA	NA	NA
Television	0	Baseline	205.0	158.00	362.50	NA	NA	NA	NA	NA
Clothes Washer	1	No Warm Rinse	814.0	190.00	380.04	0.00	0.00	1.29	111.1	111.1
Chest Freezer	0	Baseline	557.0	149.00	347.02	0.00	0.00	0.25	4.2	115.0
Upright Freezer	0	Baseline	797.5	159.00	370.31	0.00	0.00	0.03	0.4	115.0
Refrigerator	0	Baseline	955.0	224.00	521.70	0.00	0.00	1.12	128.1	243.8
Refrigerator	1	Enhanced Heat Transfer	936.0	224.10	521.94	0.11	0.13	0.22	25.1	268.9
Refrigerator	3	Foam Door	878.0	225.55	525.10	0.46	0.55	0.67	76.6	345.0
Upright Freezer	2	4.50 Compressor	719.0	161.50	375.58	0.54	0.65	0.90	12.9	358.0
Chest Freezer	3	Foam Door	508.0	150.80	350.72	0.61	0.73	0.56	9.4	367.0
Heat Pump	4	Increase indoor coil circuits	8793.9	11.88	2050.00	0.63	0.76	6.00	51.7	419.6
Refrigerator	4	5.05 Compressor	787.0	228.95	532.33	0.67	0.80	1.05	120.2	539.8
Dishwasher	1	Improve Food Filter	779.0	157.50	310.78	0.74	0.89	0.69	35.3	575.0
Upright Freezer	4	5.05 Compressor	637.0	165.00	383.29	0.75	0.91	0.94	13.5	588.0
Central Air	4	Increase indoor coil circuits	2306.7	11.90	1730.00	0.85	1.03	4.42	126.1	714.7
Electric Hot Water Heater	3	2.0" Foam + Heat Trap	4116.2	90.00	284.00	0.90	1.08	0.47	19.5	734.2
Electric Clothes Dryer	2	Moisture Term	933.0	163.00	318.69	1.14	1.37	0.63	36.5	770.0
Chest Freezer	4	5.05 Compressor	462.0	154.30	357.99	1.27	1.52	0.53	8.8	779.0
Pool Controls	1	Improve Pool Pump Controls	1600.0	50.00	100.00	1.38	1.65	10.35	9.6	789.0
Television	3	Improve CRT	171.0	161.75	369.90	2.32	2.78	0.06	4.5	793.6
Room Air	4	10.0 EER compressor	496.8	10.00	400.00	2.36	2.83	0.84	27.0	820.6
Television	1	Standby Power 2W	184.0	160.15	366.80	2.37	2.85	0.24	18.8	839.0

Table 5 (continued)
Residential Conservation Supply Curves 2000 – Electric Appliances

Appliance	Level	Design	UEC (kWh/y)	SEER	Consumer Price	Technology Only CCE ($/MMBtu)	Technology & Program CCE ($/MMBtu)	Per Unit Energy Savings (MMBtu/y)	Total Energy Savings (TBtu/y)	Cum. Energy Savings (TBtu/y)
Upright Freezer	5	2" Door	611.0	168.70	391.49	2.53	3.04	0.30	4.3	843.
Refrigerator	5	2" Door	763.0	232.65	540.63	2.91	3.49	0.28	31.7	875.3
Upright Freezer	11	Evacuated Panels	423.0	209.10	477.28	3.02	3.63	1.37	19.6	894.9
Television	2	Reduce Screen Power 5%	176.0	161.45	368.90	3.04	3.65	0.09	7.2	902.
Chest Freezer	11	Evacuated Panels	250.0	196.30	445.92	3.06	3.67	1.67	27.9	929.
Lighting	1	High Efficiency Incandescent	860.0		9.68	3.30	3.96	1.61	136.2	1066.
Heat Pump	5	Increase ou)oor coil area	8497.1	12.11	2140.00	3.32	3.98	3.41	29.4	1095.5
Clothes Washer	2	Thermostatic Valves	761.0	203.00	399.00	3.54	4.25	0.61	52.5	1148.
Dishwasher	2	Improve Motor	761.0	161.50	316.99	3.66	4.39	0.21	10.6	1158.
Chest Freezer	7	2.5" Side Insulation	395.0	169.20	390.64	3.84	4.60	0.66	11.0	1169.
Upright Freezer	8	2.5" Door	542.0	187.50	432.47	4.15	4.99	1.09	15.6	1185.2
Chest Freezer	5	2" Door	452.0	156.70	363.40	4.34	5.21	0.12	1.9	1187.1
Clothes Washer	3	Improve Motor	747.0	207.00	405.30	4.46	5.35	0.16	13.9	1201.
Electric Clothes Dryer	3	1" Insulation	914.0	169.60	328.72	4.69	5.62	0.22	12.6	1213.
Heat Pump	3	Increase indoor coil rea	9315.5	10.13	2020.00	4.94	5.92	6.35	54.7	1268.4
Refrigerator	6	Efficient Fans	732.0	241.65	559.52	5.13	6.15	0.36	41.0	1309.3
Refrigerator	11	Evacuated Panels	577.0	287.65	656.18	5.25	6.30	1.78	204.8	1514.
Central Air	3	Increase indoor coil area	2691.2	10.20	1700.00	5.26	6.31	2.23	63.6	1577.
Upright Freezer	7	2.5" Side Insulation	557.0	185.0	427.01	5.28	6.33	0.62	8.9	1586.
Refrigerator	8	3.0" Side Insulation	690.0	253.95	587.12	5.35	6.42	0.18	21.1	1607.6
Refrigerator	7	2.5" Side Insulation	706.0	249.10	576.95	5.64	6.77	0.30	34.3	1642.0
Dishwasher	3	Improve Fill Control	742.0	169.50	329.56	7.02	8.42	0.22	11.2	1653.
Electric Clothes Dryer	4	Recycle Exhaust	859.0	199.60	374.62	7.41	8.89	0.63	36.5	1689.
Electric Hot Water	4	Heat Recover Design	3300.0		900.00	7.85	9.42	9.39	389.3	2078.9
Room Air	5	Increase evaporator area	487.1	10.20	409.00	8.82	10.59	0.11	3.6	2082.5
Refrigerator	9	Two-Compressor System	607.0	303.95	692.07	10.64	12.77	0.95	109.6	2192.
Refrigerator	12	Two-Compressor System	508.0	337.65	760.72	12.75	15.30	0.79	91.1	2283.
Refrigerator	10	Adaptive Defrost	586.0	319.95	725.42	13.36	16.03	0.24	27.7	2311.
Central Air	6	Increase outdoor coil tubes	2210.1	12.42	1890.00	14.50	17.40	0.61	17.3	2328.4
Refrigerator	13	Adaptive Defrost	490.0	353.65	794.04	15.57	18.69	0.21	23.8	2352.2
Heat Pump	6	Increase outdoor coil tubes	8710.2	12.42	2200.00	19.60	23.52	0.96	8.3	2360.5
Central Air	5	Increase outdoor coil area	2263.0	12.13	1820.00	22.53	27.03	0.50	14.3	2374.8
Clothes Washer	4	Plastic Tub	743.0	213.00	414.83	23.59	28.31	0.05	4.0	2378.8

Figure 6
Conservation Supply Curve
Residential Energy 2000

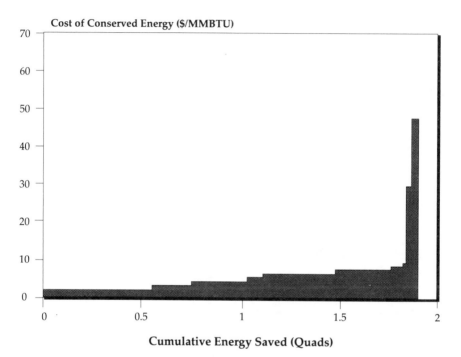

Cost of Conserved Energy ($/MMBTU)

Cumulative Energy Saved (Quads)

Shell Retrofit Electrically Heated Homes

Table 6
Residential Conservation Supply – Building Shell Improvements 2000
Electrically Heated and Cooled Houses

(Analysis based on Washington, D.C.)

	Level	Retrofit	Retrofit Cost $	Heating Energy kWh	Cooling Energy kWh	UEC Energy MMBtu	Energy Saved MMBtu/Y	CCE $/MMBtu	Total Saved TBtu/Y	Cum. Saved TBtu/Y
Heat Pump	1	Baseline	0.00	6741.559	2067.841	101.31	NA	NA	NA	NA
Resistance	1	Baseline	0.00	12467.84	2118.147	156.90	NA	NA	NA	NA
Resistance	2	Floor, from no insulation to R-19	980.00	10274.34	2020.800	131.05	25.85	3.06	314.2	314.2
Resistance	3	Infiltration, from 0.7 to 0.4 ACH	892.00	8379.228	1868.912	108.29	22.76	3.16	276.7	590.8
Resistance	4	Ceiling, from R-12.2 to R-31.2	713.00	7072.383	1812.990	92.90	15.39	3.73	187.0	777.8
Resistance	5	Walls, from R-5.6 to R-11	547.00	6194.214	1753.616	82.43	10.48	4.21	127.4	905.2
Resistance	6	Windows, from 2 to 3 panes	526.00	5403.168	1692.170	72.94	9.49	4.47	115.3	1020.5
Heat Pump	2	Floor, from no insulation to R-19	980.00	5535.670	1972.806	86.35	14.96	5.28	91.5	1112.1
Heat Pump	4	Infiltration, from 0.7 to 0.4 ACH	892.00	3791.259	1769.931	63.95	12.48	5.76	86.7	1198.9
Heat Pump	3	Ceiling, from R-12.2 to R-31.2	713.00	4822.012	1824.526	76.44	9.91	5.80	245.1	1443.8
Resistance	7	Ceiling, from R-31.2 to R-42.2	267.00	5123.295	1659.031	69.51	3.43	6.27	41.7	1485.5
Heat Pump	5	Walls, from R-5.6 to R-11	547.00	3312.143	1711.967	57.78	6.18	7.14	239.0	1724.5
Heat Pump	6	Windows, from 2 to 3 panes	526.00	2883.949	1651.981	52.16	5.61	7.55	57.1	1781.6
Heat Pump	7	Ceiling, from R-31.2 to R-42.2	267.00	2732.730	1619.629	50.05	2.11	10.19	11.2	1792.8
Resistance	8	Ceiling, from R-42.2 to R-50.2	535.00	4995.309	1643.152	67.93	1.57	27.41	19.1	1811.9
Heat Pump	8	Ceiling, from R-42.2 to R-50.2	535.00	2663.293	1604.127	49.08	0.98	44.14	5.1	1817.0

Heat Pump (9.46 SEER, 7.61 HSPF)
Central Air (11.9 SEER)

Figure 7
Conservation Supply Curve
Residential Energy 2000

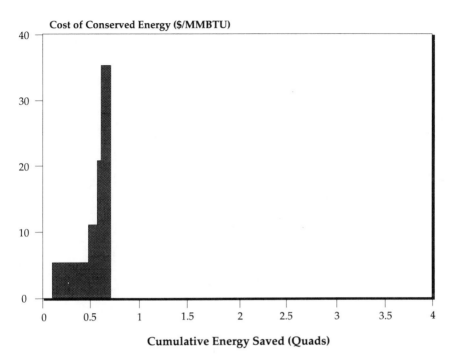

Cumulative Energy Saved (Quads)

Gas Appliances Only

326

Table 7
Residential Conservation Supply Curves 2000
Gas Appliances

Appliance	Level	Design	UEC (kWh/y)	SEER	Consumer Price	Technology Only CCE ($/MMBtu)	Technology & Program CCE ($/MMBtu)	Per Unit Energy Savings (MMBtu/y)	Total Energy Savings (TBtu/y)	Cum. Energy Savings (TBtu/y)
Gas Furnace	0	Baseline	57.80	76.50	2665.50	NA	NA	NA	NA	NA
Gas Hot Water	0	Baseline	17.50		360.00	NA	NA	NA	NA	NA
Gas Clothes Dryer	0	Baseline	4.00	170.00	340.03	NA	NA	NA	NA	NA
Gas Furnace	3	Induced Draft	55.30	80.00	2669.0	0.12	0.15	2.5	116.7	116.7
Gas Clothes Dryer	2	Moisture Term	3.52	183.00	358.70	3.62	4.35	0.2	3.2	119.8
Gas Clothes Dryer	1	Temperature Term	3.72	178.00	351.60	4.22	5.06	0.3	4.4	124.3
Gas Furnace	5	Condensing Heat Exchanger	48.10	92.00	3040.00	4.54	5.44	6.9	316.7	441.0
Gas Furnace	4	Insulation	54.90	80.50	2689.00	5.17	6.20	0.3	15.8	456.9
Gas Hot Water	5	2.0" Foam + IID	15.80	70.00	510.00	10.44	12.52	1.7	91.3	548.1
Gas Clothes Dryer	3	1" Insulation	3.45	189.60	368.72	14.62	17.54	0.1	1.1	549.2
Gas Hot Water	6	Multiple Flues	15.10	73.00	610.0	18.46	22.15	0.6	34.4	583.6
Gas Clothes Dryer	4	Recycle Exhaust	3.24	219.60	414.63	22.32	26.79	0.2	3.3	587.0
Gas Hot Water	7	Pulse Condensing	13.50	82.00	1010.00	28.83	34.60	1.7	88.1	675.1

Figure 8
Conservation Supply Curve
Residential Energy 2000

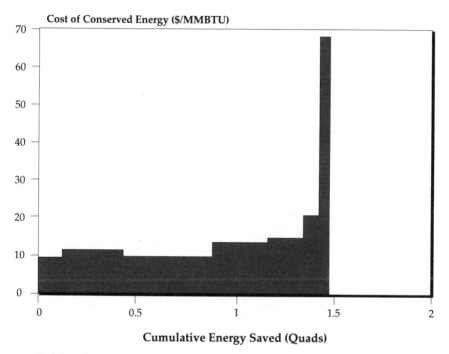

Cost of Conserved Energy ($/MMBTU)

Cumulative Energy Saved (Quads)

Shell Retrofit Gas Heated Homes

Table 8
Residential Conservation Supply – Building Shell Improvements 2000
Gas Heated Homes

(Analysis based on Washington, D.C.)

Level	RETROFIT	Retrofit Cost $	Heating Energy Therms	Cooling Energy kWh	Energy MMBtu	Gas Energy MMBtu	Elec. Energy MMBtu	Energy Saved MMBtu/Y	Gas Saved MMBtu/Y	Elec. Saved MMBtu/Y
Gas Heat 1	Baseline	0.00	556.4701	2322.675	82.36	55.65	26.71	NA	NA	NA
Gas Heat 2	Floor, from no insul. to R-19	980.00	458.4355	2216.842	71.34	45.84	25.49	9.13	8.16	0.97
Gas Heat 3	Ceiling, from R-12.2 to R-31.2	713.00	399.8968	2051.225	63.58	39.99	23.59	6.39	4.87	1.52
Gas Heat 4	Infilt., from 0.7 to 0.4 ACH	892.00	315.2625	1990.633	54.42	31.53	22.89	7.60	7.04	0.56
Gas Heat 5	Walls, from R-5.6 to R-11	547.00	276.4718	1926.002	49.80	27.65	22.15	3.82	3.23	0.59
Gas Heat 6	Windows, from 2 to 3 panes	526.00	241.2076	1858.948	45.50	24.12	21.38	3.55	2.93	0.62
Gas Heat 7	Ceiling, R-42.2	267.00	228.5124	1823.401	43.82	22.85	20.97	1.38	1.06	0.33
Gas Heat 8	Ceiling, R-50.2	535.00	221.4596	1802.395	42.87	22.15	20.73	0.78	0.59	0.19

Level	RETROFIT	Energy CCE $/MMBtu	Gas CCE $/MMBtu	Elec. CCE $/MMBtu	Total Energy TBtu	Total Gas TBtu	Total Elec. TBtu	Cum. Energy TBtu	Cum. Gas TBtu	Cum. Elec. TBtu
Gas Heat 1	Baseline	NA	NA	NA	NA	NA	NA	NA	NA	NA
Gas Heat 2	Floor, from no insul. to R-19	8.65	9.68	81.11	169.1	134.3	34.8	169.1	134.3	34.8
Gas Heat 3	Ceiling, from R-12.2 to R-31.2	8.99	11.80	37.71	367.7	313.3	54.4	536.8	447.6	89.2
Gas Heat 4	Infilt., from 0.7 to 0.4 ACH	9.46	10.21	128.95	472.9	453.0	19.9	1009.7	900.6	109.1
Gas Heat 5	Walls, from R-5.6 to R-11	11.53	13.66	74.13	228.8	207.6	21.2	1238.5	1108.2	130.3
Gas Heat 6	Windows, from 2 to 3 panes	11.94	14.45	68.71	210.8	188.7	22.0	1449.3	1297.0	152.3
Gas Heat 7	Ceiling, R-42.2	15.55	20.37	65.79	79.6	67.9	11.7	1528.9	1364.9	164.0
Gas Heat 8	Ceiling, R-50.2	55.27	73.47	223.10	44.6	37.7	6.9	1573.5	1402.7	170.9

Figure 9
Conservation Supply Curve
Residential Energy 2010

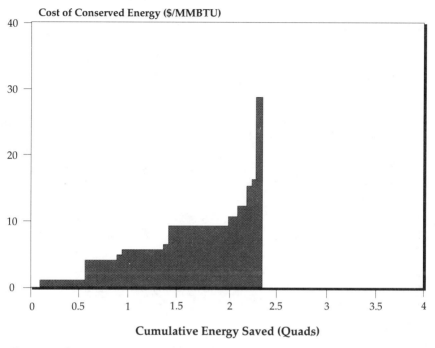

Cost of Conserved Energy ($/MMBTU)

Cumulative Energy Saved (Quads)

Electric Appliances Only

Table 9
Residential Conservation Supply Curves 2010 – Electric Appliances

Appliance	Level	Design	UEC (kWh/y)	SEER	Consumer Price	Technology Only CCE ($/MMBtu)	Technology & Program CCE ($/MMBtu)	Per Unit Energy Savings (MMBtu/y)	Total Energy Savings (TBtu/y)	Cum. Energy Savings (TBtu/y)
Heat Pump	0	Baseline	9827.4	9.52	1788.37	NA	NA	NA	NA	NA
Central Air Conditioner	0	Baseline	2868.3	9.57	1554.17	NA	NA	NA	NA	NA
Lighting	0	Baseline	1000.0		4.72	NA	NA	NA	NA	NA
Color Television	0	Baseline	205.0	158.00	362.50	NA	NA	NA	NA	NA
Electric Water Heater	0	Baseline	4130.0	89.70	282.80	NA	NA	NA	NA	NA
Room Air Conditioner	0	Baseline	557.6	8.91	379.76	NA	NA	NA	NA	NA
Clothes Washer	0	Baseline	911.6	190.00	380.04	NA	NA	NA	NA	NA
Pool Controls	0	Baseline	2500.0	0.00	0.00	NA	NA	NA	NA	NA
Chest Freezers	0	Baseline	477.7	153.20	353.25	NA	NA	NA	NA	NA
Dishwasher	0	Baseline	821.5	150.00	307.74	NA	NA	NA	NA	NA
Refrigerator	0	Baseline	931.0		522.23	NA	NA	NA	NA	NA
Upright Freezers	0	Baseline	660.4		380.67	NA	NA	NA	NA	NA
Electric Clothes Dryer	0	Baseline	944.0	159.00	313.00	NA	NA	NA	NA	NA
Clothes Washer	1	No Warm Rinse	814.0	190.00	380.04	0.00	0.00	1.12	108.7	108.7
Refrigerator	3	Foam Door	878.0	225.55	525.10	0.46	0.55	0.61	76.7	185.4
Heat Pump	4	Increase indoor coil circuits	8793.9	11.88	2050.00	0.63	0.76	6.00	60.4	245.8
Refrigerator	4	5.05 Compressor	787.0	228.95	532.33	0.67	0.80	1.05	131.7	377.5
Dishwasher	1	Improve Food Filter	779.0	157.50	310.78	0.76	0.91	0.49	29.1	406.6
Central Air Conditioner	4	Increase indoor circuits	2306.7	11.90	1730.00	0.85	1.03	4.42	153.9	560.5
Upright Freezers	4	5.05 Compressor	637.0	165.00	383.29	0.90	1.08	0.27	4.1	564.7
Electric Water Heater	3	2.0" Foam + Heat Trap	4116.2	90.00	284.00	0.91	1.09	0.16	7.5	572.2
Pool Controls	1	Improve Controls	1600.0	50.00	100.00	1.38	1.65	10.35	10.5	582.6
Color Television	3	Improve CRT	171.0	161.75	369.90	2.32	2.78	0.06	0.5	583.1
Color Television	1	Standby Power 2W	184.0	160.15	366.80	2.37	2.85	0.24	2.0	585.2
Chest Freezers	4	5.05 Compressor	462.0	154.30	357.99	2.43	2.91	0.18	3.2	588.4
Upright Freezers	5	2" Door	611.0	168.70	391.49	2.53	3.04	0.30	4.6	593.0
Refrigerator	5	2" Door	763.0	232.65	540.63	2.91	3.49	0.28	34.7	627.7
Upright Freezers	11	Evacuated Panels	423.0	209.10	477.28	3.02	3.63	1.37	21.0	548.7
Color Television	2	Reduce Screen Power 5%	176.0	161.45	368.90	3.04	3.65	0.09	0.8	549.5
Room Air Conditioner	4	10.0 EER compressor	496.8	10.00	400.00	3.18	3.82	0.70	25.5	675.0

Table 9 (continued)
Residential Conservation Supply Curves 2010 – Electric Appliances

Appliance	Level	Design	UEC (kWh/y)	SEER	Consumer Price	Technology Only CCE ($/MMBtu)	Technology & Program CCE ($/MMBtu)	Per Unit Energy Savings (MMBtu/y)	Total Energy Savings (TBtu/y)	Cum. Energy Savings (TBtu/y)
Chest Freezers	11	Evacuated Panels	250.0	196.30	445.92	3.28	3.93	2.32	41.6	716.6
Lighting	1	High Efficiency Incandescent	860.0		9.68	3.30	3.96	1.61	148.1	864.7
Heat Pump	5	Increase outdoor coil area	8497.1	12.11	2140.00	3.32	3.98	3.41	34.4	899.1
Clothes Washer	2	Thermostatic Valves	761.0	203.00	399.00	3.54	4.25	0.61	59.1	958.2
Dishwasher	2	Improve Motor	761.0	161.50	316.99	3.66	4.39	0.21	12.3	970.5
Chest Freezers	7	2.5" Side Insulation	395.0	169.20	390.64	3.84	4.60	0.66	11.7	982.2
Upright Freezers	8	2.5" Door	542.0	187.50	432.47	4.15	4.99	1.09	16.8	999.0
Chest Freezers	5	2" Door	452.0	156.70	363.40	4.34	5.21	0.12	2.1	1001.0
Clothes Washer	3	Improve Motor	747.0	207.00	405.30	4.46	5.35	0.16	15.6	1016.6
Electric Clothes Dryer	2	Moisture Term	933.0	163.00	318.69	4.59	5.51	0.13	8.3	1025.0
Electric Clothes Dryer	3	1" Insulation	914.0	169.60	328.72	4.69	5.62	0.22	14.4	1039.3
Heat Pump	3	Increase indoor coil area	9315.5	10.13	2020.00	4.95	5.94	5.89	59.3	1098.6
Refrigerator	6	Efficient Fans	732.0	241.65	559.52	5.13	6.15	0.36	44.9	1143.5
Refrigerator	11	Evacuated Panels	577.0	287.65	656.18	5.25	6.30	1.78	224.3	1367.8
Upright Freezers	7	2.5" Side Insulation	557.0	185.00	427.01	5.28	6.33	0.62	9.5	1377.3
Refrigerator	8	3.0" Side Insulation	690.0	253.95	587.12	5.35	6.42	0.18	23.2	1400.4
Refrigerator	7	2.5" Side Insulation	706.0	249.10	576.95	5.64	6.77	0.30	37.6	1438.1
Dishwasher	3	Improve Fill Control	742.0	169.50	329.56	7.02	8.42	0.22	13.0	1451.1
Electric Clothes Dryer	4	Recycle Exhaust	859.0	199.60	374.62	7.41	8.89	0.63	41.6	1492.7
Electric Water Heater	4	Heat Recovery	3300.0		900.00	7.85	9.42	9.39	445.6	1938.4
Room Air Conditioner	5	Increase evaporator area	487.1	10.20	409.00	8.82	10.59	0.11	4.1	1942.4
Central Air Conditioner	3	Increase indoor coil area	2691.2	10.20	1700.00	9.01	10.81	2.04	70.9	2013.4
Refrigerator	9	Two-Compressor System	607.0	303.95	692.07	10.64	12.77	0.95	120.1	2133.5
Refrigerator	12	Two-Compressor System	508.0	337.65	760.72	12.75	15.30	0.79	99.8	2233.3
Refrigerator	10	Adaptive Defrost	586.0	319.95	725.42	13.36	16.03	0.24	30.4	2263.7
Central Air Conditioner	6	Increase outdoor coil tubes	2210.1	12.42	1890.00	14.49	17.38	0.61	21.2	2284.9
Refrigerator	13	Adaptive Defrost	490.0	353.65	794.04	15.57	18.69	0.21	26.0	2310.9
Heat Pump	6	Increase outdoor coil tubes	8710.2	12.42	2200.00	19.60	23.52	0.96	9.7	2320.6
Central Air Conditioner	5	Increase outdoor coil area	2263.0	12.13	1820.00	22.55	27.06	0.50	17.5	2338.1
Clothes Washer	4	Plastic Tub	743.0	213.00	414.83	23.59	28.31	0.05	4.5	2342.6

Figure 10
Conservation Supply Curve
Residential Energy 2010

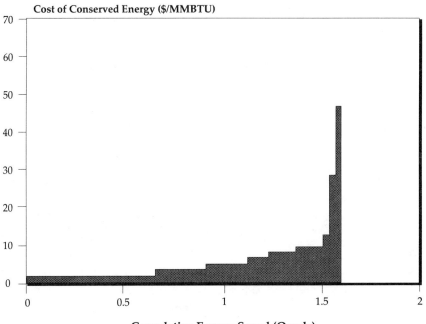

Shell Retrofit Electrically Heated Homes

Table 10

Residential Conservation Supply – Building Shell Improvements 2010 Electrically Heated and Cooled Homes

(Analysis based on Washington, D.C.)

	Level	Retrofit	Retrofit Cost $	Heating Energy kWh	Cooling Energy kWh	UEC MMBtu	Energy Saved MMBtu/Y	CCE $/MMBtu	Total Energy TBtu/Y	Cum. Energy TBtu/Y
Heat Pump	1	Baseline	0.00	6681.19	2067.69	100.61	NA	NA	NA	NA
Resistance	1	Baseline	0.00	12765.63	1824.73	158.45	NA	NA	NA	NA
Resistance	2	Floor, from no insulation to R-19	980.00	10519.75	1741.59	132.09	26.36	3.00	349.1	349.1
Resistance	3	Ceiling, from R-12.2 to R-31.2	892.00	8579.36	1697.16	109.49	22.60	3.18	299.3	648.5
Resistance	4	Infiltration, from 0.7 to 0.4 ACH	713.00	7241.31	1563.88	93.26	16.24	3.54	215.1	863.6
Resistance	5	Walls, from R-5.6 to R-11	547.00	6342.16	1513.10	82.59	10.66	4.13	141.3	1004.8
Resistance	6	Windows, from 2 to 3 panes	526.00	5532.22	1460.42	72.94	9.65	4.39	127.8	1132.7
Heat Pump	2	Floor, from no insulation to R-19	980.00	5486.10	1972.66	85.78	14.84	5.32	98.9	1231.6
Heat Pump	4	Infiltration, from 0.7 to 0.4 ACH	892.00	3757.31	1769.80	63.56	12.38	5.81	82.5	1314.1
Heat Pump	3	Ceiling, from R-12.2 to R-31.2	713.00	4778.83	1824.39	75.94.	9.84	5.84	65.6	1379.7
Resistance	7	Ceiling, from R-31.2 to R-42.2	267.00	5245.66	1432.49	69.47	3.47	6.19	46.0	1425.7
Heat Pump	5	Walls, from R-5.6 to R-11	547.00	3282.48	1711.84	57.43	6.13	7.19	40.9	1466.6
Heat Pump	6	Windows, from 2 to 3 panes	526.00	2858.12	1651.86	51.86	5.57	7.61	37.1	1503.7
Heat Pump	7	Ceiling, from R-31.2 to R-42.2	267.00	2708.26	1619.51	49.77	2.10	10.27	14.0	1517.7
Resistance	8	Ceiling, from R-42.2 to R-50.2	535.00	5114.62	1418.53	67.87	1.60	27.01	21.1	1538.9
Heat Pump	8	Ceiling, from R-42.2 to R-50.2	535.00	2639.45	1604.01	48.80	0.97	44.46	6.5	1545.3

Heat Pump (12.11 SEER, 7.6 HSPF)
Electric Heat (11.9 SEER Air Conditioner)

Figure 11
Conservation Supply Curve
Residential Energy 2010

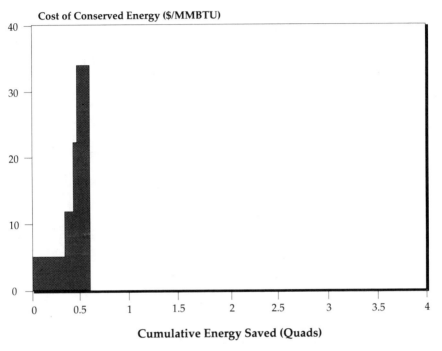

Gas Appliances Only

Table 11
Residential Conservation Supply Curves 2010
Gas Appliances

Appliance	Level	Design	UEC (kWh/y)	Consumer Price	Only CCE ($/MMBtu)	Technology & Program CCE ($/MMBtu)	Technology Per Unit Savings (MMBtu/y)	Total Savings (TBtu/y)	Cum. Savings (TBtu/y)
Gas Water Heater	0	Baseline	17.50	360.00	NA	NA	NA	NA	NA
Gas Clothes Dryer	0	Baseline	4.00	340.03	NA	NA	NA	NA	NA
Gas Furnace	0	Baseline	55.40	2710.37	NA	NA	NA	NA	NA
Gas Clothes Dryer	2	Moisture Term	3.52	358.70	3.62	4.35	0.20	3.54	3.5
Gas Clothes Dryer	1	Temperature Term	3.72	351.60	4.22	5.06	0.28	4.96	8.5
Gas Furnace	5	Condensing Heat Exchanger	48.90	3040.00	4.50	5.39	6.50	333.10	341.6
Gas Water Heater	5	2.0" Foam + IID	15.80	510.00	10.26	12.31	1.75	97.98	439.6
Gas Clothes Dryer	3	1" Insulation	3.45	368.72	14.62	17.54	0.07	1.24	440.8
Gas Water Heater	6	Multiple Flues	15.10	610.00	18.46	22.15	0.65	36.30	477.1
Gas Clothes Dryer	4	Recycle Exhuast	3.24	414.63	22.32	26.79	0.21	3.72	480.8
Gas Water Heater	7	Pulse Condensing	13.50	1010.00	28.83	34.60	1.66	92.96	573.8

Figure 12
Conservation Supply Curve
Residential Energy 2010

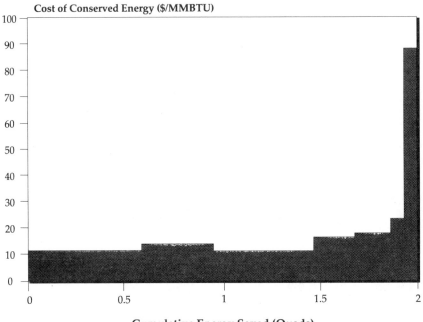

Cumulative Energy Saved (Quads)

Shell Retrofit Gas Heated Homes

Table 12
Residential Conservation Supply – Building Shell Improvements 2010 Gas Heated Homes

(Analysis based on Washington, D.C.)

	RETROFIT	Retrofit Cost $	Heating Energy Therms	Cooling Energy kWh	Energy MMBtu	Gas MMBtu	Elec. MMBtu	Energy Saved MMBtu/Y	Gas Saved MMBtu/Y	Elec. Saved MMBtu/Y
1	Baseline	0.00	483.39033	1837.13	69.47	48.34	21.13	NA	NA	NA
2	Floor, from no insulation to R-19	980.00	398.23031	1753.42	59.99	39.82	20.16	9.48	8.52	0.96
3	Ceiling, from R-12.2 to R-31.2	713.00	347.37936	1622.43	53.40	34.74	18.66	6.59	5.09	1.51
4	Infiltration, from 0.7 to 0.4 ACH	892.00	273.85992	1574.50	45.49	27.39	18.11	7.90	7.35	0.55
5	Walls, from R-5.6 to R-11	547.00	240.16351	1523.38	41.54	24.02	17.52	3.96	3.37	0.59
6	Windows, from 2 to 3 panes	526.00	209.53041	1470.34	37.86	20.95	16.91	3.67	3.06	0.61
7	Ceiling, R-42.2	267.00	198.50249	1442.23	36.44	19.85	16.59	1.43	1.10	0.32
8	Ceiling, R-50.2	535.00	192.37587	1425.61	35.63	19.24	16.39	0.80	0.61	0.19

	RETROFIT	Energy CCE $/MMBtu	Gas CCE $/MMBtu	Elec. CCE $/MMBtuTBtu	Total Energy TBtu	Total Gas TBtu	Total Elec. TBtu	Cum. Energy TBtu	Cum. Gas TBtu	Cum. Elec. TBtu
1	Baseline				645.2	597.0	48.1	645.2	597.0	48
2	Floor, from no insulation to R-19	10.00	11.13	98.45	431.8	356.5	75.3	1077.0	953.5	123
3	Ceiling, from R-12.2 to R-31.2	10.46	13.56	45.77	543.0	515.4	27.6	1620.0	1468.9	151
4	Infiltration, from 0.7 to 0.4 ACH	10.91	11.73	156.51	265.6	236.2	29.4	1885.6	1705.2	180
5	Walls, from R-5.6 to R-11	13.37	15.70	89.98	245.3	214.8	30.5	2130.9	1919.9	210
6	Windows, from 2 to 3 panes	13.85	16.60	83.40	93.5	77.3	16.2	224.3	1997.2	227
7	Ceiling, R-42.2	18.10	23.41	79.85	52.5	43.0	9.6	2276.8	2040.2	236
8	Ceiling, R-50.2	64.37	84.45	270.78						

Appendix Notes

1   *LBL's Residential Energy Model, which includes a database of elasticities, saturations, unit energy consumptions, conservation measures, capital costs, and other parameters* (Berkeley: Lawrence Berkeley Laboratory, 1991).

2   For refrigerators and freezers, all designs assume substitutes for chlorofluorocarbons. The energy performance and costs of these substitutes are taken into account. (See U.S. Department of Energy, *Technical Support Document: Energy Conservation Standards for Consumer Products: Refrigerators and Furnaces*, DOE/CE-0277, November 1989.)

3   P.M. Miller, J.H. Eto, and H.S.. Geller, *The Potential for Electricity Conservation in New York State*, American Council for an Energy-Efficient Economy for New York State Energy Research and Development Agency, September 1989.

4   H. Ruderman, M.D. Levine, and J.E. McMahon, "The Behavior of the Market for Energy Efficiency in Residential Appliances Including Heating and Cooling Equipment," in *Energy Systems and Policy*, volume 10, 1987.

5   U.S. Energy Information Administration, S*tate Energy Prices for Residential Sector*, 1989-1990, EAFD/89-02, July 1989.

Bibliography

A. Meier, J. Wright, and A.H. Rosenfeld, *Supplying Energy Through Greater Efficiency*, University of California Press, Berkeley and Los Angeles, California (1983).

Northwest Power Planning Council, "Northwest Conservation and Electric Power Plan," Portland, Oregon (1986).

The Michigan Electricity Options Study, Department of Commerce, State of Michigan, 1988.

D.B. Pirkey and R.M. Scheer, "Energy Conservation Potential: A Review of Eight Studies," *Proceedings of the ACEEE 1988 Summer Study on Energy Efficiency in Buildings*, Asilomar, California (1988).

A. Meier, J. Wright, and A. Rosenfeld, "Supply Curves of Conserved Energy For California's Residential Sector," *Energy—The International Journal* 7 (1982): 347–58.

A. Meier, "Supply Curves of Conserved Energy," Ph.D. dissertation, Energy and Resources Program, University of California, Berkeley, 1982.

Appendix 4

Six States' Energy Usage by End-Use Sector, 1988
(in trillions of Btu's and percentage of total)

State	Industrial	Transportation	Residential	Commercial	Total
Florida	432	1110	750	636	2928
	15%	38%	26%	22%	100%
Iowa	348	232	223	139	942
	37%	25%	24%	15%	100%
Nevada	84	133	75	63	355
	24%	37%	21%	18%	100%
New York	692	874	1025	995	3586
	19%	24%	29%	28%	100%
Washington	640	513	359	294	1806
	35%	28%	20%	16%	100%
Wisconsin	457	341	364	230	1392
	33%	24%	26%	17%	100%

Appendix 5
The 1990 Iowa Comprehensive Energy Plan

The 1990 Iowa Comprehensive Energy Plan is summarized below. It is the first in a series of energy plans to be completed every other year at the directive of the governor and the Iowa General Assembly. Incorporating research and discussion from private industry, public interest groups, and state and local government planners, it examines Iowa's consumption of energy and the economic and environmental impacts of energy use. The energy plan seeks to improve Iowa's energy planning by meeting future energy demand with increased efficiency rather than new supply and by increasing the use of alternative energy sources until they make up 10 percent of the total energy mix in Iowa. The following ten specific recommendations are made as part of the plan.

1) Make it easier to use electricity. Iowa's utilities are urged to implement the following principles as part of a least-cost planning strategy: aggressively implement demand-side management, decouple utility profits from sales, and incorporate "externalities" or social costs into utility planning.

2) Improve efficiency of the state's transportation sector. Currently, the taxes collected for road funding depend on gas sales revenue. As a result, the state has no incentive to promote fuel efficiency. The plan calls for a change in this structure. To promote transportation efficiency, the plan recommends providing incentives for individuals to drive fuel efficient cars, increasing the state government's automobile fleet fuel efficiency, and studying and promoting telecommuting as a commuting alternative.

3) Make Iowa's buildings more efficient energy-users. Iowa's energy plan calls for a comprehensive effort to make buildings more energy efficient. Recommendations include developing a statewide energy rating system, creating market incentives to promote the use of the rating system, providing exemptions for energy efficiency improvements from property taxes, and encouraging the use of utility connection bonuses and assessments based on new buildings' energy efficiency.

4) Improve the efficiency of residences to make them more affordable. To increase residential energy efficiency, the following policy suggestions are put forth: increase the Iowa Weatherization Program, promote mortgage guidelines that encourage energy efficient homes, continue home energy audits, and give the Utilities Board the authority to require utilities to arrange for the financing of energy efficiency improvements for their consumers.

5) Increase the energy efficiency of agricultural operations and encourage the production of energy crops. The energy plan recommends that the agricultural sector's energy policy include assisting farmers to lessen their environmental impacts and to reduce their dependence on nonrenewable energy resources, promoting agriculture's potential to produce renewable energy sources, and reforesting of marginal cropland.

6) Encourage the development of alternative energy production as an effort to diversify Iowa's energy resource base. Iowa's university system is expected to step up its research of alternative-fueled vehicles and wind and solar research.

7) Incorporate external costs into future energy policy decisions. Iowa is expected to join other states in the quantification of both negative and positive externalities that result from electrical power production.

8) Develop, commercialize, and promote energy efficiency and alternative energy technologies and businesses. Business recruiting should include the targeting of companies and industries that focus on recycling, pollution prevention, pollution abatement, alternative fuels, and energy efficiency.

9) Educate key groups on energy efficiency. Homeowners, homebuilders, the agricultural community, and business, industry and local government leaders need to receive information to enhance their understanding and commitment to energy efficiency and alternative resources.

10) Develop better energy data collection, forecasting, and evaluation techniques and programs for state government energy planning. Iowa's Department of Natural Resources (DNR) will increase its role in the state energy planning process through the 1990 comprehensive energy plan. The DNR is expected to devote more attention to evaluations of programs in place and collecting more energy-saving statistics from programs yet to be implemented.

Appendix 6
Nevada Senate Bill 497

SENATE BILL NO. 497—SENATORS GETTO, TOWNSEND, BEYER,
HICKEY, HORN, JOERG, O'DONNELL, SMITH AND VERGIELS

May 26, 1989

Referred to Committee on Commerce and Labor

SUMMARY—Revises requirements for plan submitted by electrical utility to
increase supply and reduce demand. (BDR 58-1717)

FISCAL NOTE: Effect on Local Government: No.
                      Effect on the State or on Industrial Insurance: No.

EXPLANATION—Matter in *italics* is new; matter in brackets
[ ] is material to be omitted.

AN ACT relating to public utilities; requiring that the plan submitted by an
electrical utility to increase the supply of or reduce the demand for electricity
must minimize expenditures by the utility and demonstrate the economic and
environmental benefits to be realized through the use of certain efficiency
measures and sources of supply; and providing other matters properly
relating thereto.

THE PEOPLE OF THE STATE OF NEVADA, REPRESENTED IN SENATE
AND ASSEMBLY, DO ENACT AS FOLLOWS:

Section 1, NRS 704.746 is hereby amended to read as follows:

704.746 1. Not more than 60 days after a utility has filed its plan, the
commission shall convene a public hearing on the adequacy of the plan.

2. At the hearing any interested person may make comments to the
commission regarding the contents and adequacy of the plan.

3. After the hearing the commission shall determine whether:
(a) The [utility's] forecast requirements *of the utility* are based on
substantially accurate data and an adequate method of forecasting. [;]
(b) The plan identifies and takes into account any present and
projected reductions in the demand for energy [which] *that* may result
from measures [for conservation and management of loads] *to
improve energy efficiency* in the industrial, commercial, residential
and energy producing sectors of the area being served. [; and

(c) The utility's plan shows an adequate consideration of]
(c) The plan minimizes total expenditures by the utility while maintaining levels of service that are adequate and reliable.
(d) The plan explicitly calculates and shows the economic and environmental benefits to this state associated with the following possible measures and sources of supply:

    (1) [Conservation;

    (2) Load management;

    (3)] *Improvements in energy efficiency;*

    (2) Pooling of power;

    [(4)] (3) Purchases of power from neighboring states [or counties;

    (5) Facilities which]

    (4) *Facilities that* operate on solar or geothermal energy or wind; and

    (5) *Facilities that* operate on the principle of cogeneration or hydrogeneration.

(e) The plan gives preference over other sources of supply to the measures and sources of supply set forth in paragraph (d) that provide the greatest economic and environmental benefits to the state and are consistent with minimizing total expenditures by the utility while providing levels of service that are adequate and reliable. The commission shall determine the level of preference to be given to those measures and sources of supply.

Sec. 2. This act becomes effective upon passage and approval.

Appendix 7
Selected Utility Demand-Side Management Budgets
(As a percentage of revenue)

Utility	1990 Total DSM Budget (Million $)	1990 DSM Budget Percentage of Revenue	1990 Projected Energy Impact (GWh)	1990 Projected Peak Impact (MW)
Bonneville Power Administration	108	6.0	3330	35
New England Electric System	65	4.0	130	160
Central Maine Power	26	3.7	130	30
Wisconsin Electric	42	3.5	200	50
Puget Sound Power & Light	24	2.7	70	0
Pacific Gas & Electric	138	2.6	370	120
Florida Power	40	2.5	35	650*
Boston Edison	26	2.4	30	5
Nevada Power Company (1)	5	1.0	35	11
Sierra Pacific Power Company (1)	2	0.6	6	12

Notes: *Cumulative Megawatt (MW) Reduction.
(1) Numbers are actual for 1990.

Source: Cambridge Energy Research Associates, *Lightening the Load: Electric Utilities and Demand-Side Management*, p. 12, Nevada Power Company.
Sierra Pacific Power Company, *Progress Report on Three-Year Action Plan*, 1991.

Appendix 8
Washington State Energy-Efficiency Measures

Key Legislative Dates

1977   Washington state adopts its first statewide mandatory standard for energy efficiency in new home construction.

1980   The United States Congress passes the Pacific Northwest Power Planning Conservation Act (Northwest Power Act). The act creates a state-appointed Northwest Power Planning Council (NWPPC), which makes judgments about the region's future electrical energy demand and resources, including conservation, and develops the Northwest Power Plan. According to the act, Bonneville Power Administration's (BPA) resource acquisition strategies must be consistent with the Northwest Power Plan.

1980   Washington state upgrades standards for energy efficiency in new home construction.

1983   As required by the Northwest Power Plan, the NWPPC publishes model conservation standards (MCS) for both residential and commercial structures and urges that they become the energy code regionwide for electrically heated structures.

1983   BPA begins offering a regional code adoption assistance program, originally known as the Early Adopter Program. The Program offers financial incentives to local MCS. (Later, the program will be called the Northwest Energy Code Program.)

1984   Tacoma becomes the first jurisdiction in the Pacific Northwest to enforce the Model Conservation Standards.

1985   The Northwest's publicly owned electric utilities, in cooperation with BPA, begin sponsoring the Super Good Cents (SGC) program. This marketing effort paves the way for jurisdictions to adopt MCS-equivalent energy codes by offering financial incentives to builders who build electrically heated homes to a higher level of energy efficiency. Privately owned utilities begin offering their own programs, such as Comfort Plus.

1986   Washington state codifies the 1986 Energy Code, which achieves 60 percent of the efficiency level required by the MCS. (The 1986 Washington State Energy Code remained in effect until the 1991 Energy Code went into effect July 1, 1991.)

1987   Bonneville produces the Northwest Energy Code (NWEC), a prototype code that jurisdictions can adopt by reference. Based on performance and cost data taken from 500 homes built according to the model conservation standards, the NWEC does not require air-to-air heat exchangers and air vapor barriers.

1987   BPA's Early Adopter Program is renamed the NWEC Program. It encourages local jurisdictions to adopt the prototype NWEC as a local code.

1990   The 1990 energy conservation legislation–Engrossed Substitute House Bill (ESHB) 2198 – is passed by the Washington State Legislature and signed by the governor. The legislation enacts new standards for energy efficiency in residential construction. The insulation levels established for electrically heated homes are very similar to those in the NWEC.

1991   The standards set forth in ESHB 2198 are published in code format.

1991   Each local government in Washington state enforces the new code as of July 1, 1991, unless the jurisdiction by March 1, 1990, had adopted a local code that exceeds ESHB 2198's requirements.

Understanding ESHB 2198

In January, 1990 the Washington State Legislature enacted new standards for energy efficiency in residential construction. The legislation, ESHB 2198, was signed into law by Governor Booth Gardner on February 6, 1990. Technically, the legislation, which is identified as "an act relating to energy efficiency and conservation," does not constitute a new state-wide energy code. The legislation sets forth the standards that the State Building Code Council must use to prepare a new Washington State Energy Code (WSEC).

New Residential Construction Efficiency Standards

1) Requires new residential construction to meet a certain level of energy efficiency while providing flexibility in the actual building design.

2) Requires that towns, cities, and counties begin enforcing the WSEC July 1, 1991 unless they are covered under the Northwest Energy Code alternative.

3) Stipulates that the code will take into account regional climatic conditions.

4) Specifies that the code will apply to new residential construction and sets standards for various heating sources.

6) Establishes in the state treasury a training account. These funds will be used to foster accurate inspections and proper enforcement of the WSEC. The legislation stipulates how revenues for this account will be generated.

7) Requires electric utilities to make payments of $900 per house for houses that are less than 2,000 square feet and $390 per unit of multi-family structures. Payments are made to owners at time of construction.

8) Payments are required for residential construction occurring between July 1, 1991, and July 1, 1995, that uses electric resistance space heat.

9) Indicates that the energy code for new non-residential buildings continues to be the 1986 amended edition of the Washington State Energy Code.

Source: Washington State Energy Office, 1991.

Appendix 9
Washington State Industrial Buildings Program Energy Conservation

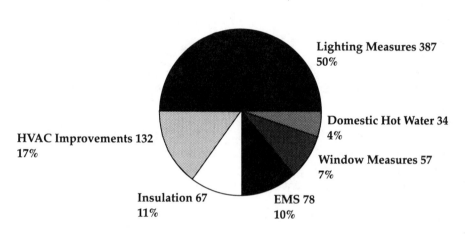

ECM Breakdown by Frequency

Lighting Measures 387
50%

Domestic Hot Water 34
4%

Window Measures 57
7%

HVAC Improvements 132
17%

Insulation 67
11%

EMS 78
10%

Source: *Institutional Buildings Program in Washington State,* Washington State Energy Office, December 1987.

Appendix 10
State Energy Office Contacts

K. David Shropshire, Division Chief
Science, Technology and Energy
    Division
401 Adams Avenue
P.O. Box 5690
Montgomery, Alabama 36103-5690
205-242-5292

Steve Baden, Chief
Conservation Section
Rural Development Division
333 West 4th Avenue, Suite 220
Anchorage, Alaska 99501
907-269-4632

Matt T. Le'i, Director
Territorial Energy Office
American Samoa Government
Samoa Energy House, Tafuna
Pago Pago, American Samoa 96799
684-699-1101/2/3/4

Jack Haenichen, Director
Arizona Energy Office
Department of Commerce
3800 North Central Avenue
Phoenix, Arizona 85013
602-280-1402

Jim Blakely, Director
Energy Division
Arkansas Industrial Commission
No. 1 State Capitol Mall
Little Rock, Arkansas 72201
501-682-7315

Charles R. Imbrecht, Chair
California Energy Commission
1516 Ninth Street
Sacramento, California 95814
916-654-4001

Nancy Tipton, Acting Director
Colorado Office of Energy
    Conservation
1675 Broadway, Suite 1300
Denver, Colorado 80202-4613
303-620-4292

Susan Shimelman
Under Secretary for Energy
Office of Policy and Management
80 Washington Street
Hartford, Connecticut 06106
203-566-4298

George P. Donnelly, Director
Division of Facility Management
P.O. Box 1401
O'Neill Building
Dover, Delaware 19903
302-739-5644

Charles J. Clinton, Director
D.C. Energy Office
613 G Street, N.W., 5th Floor
Washington, D.C. 20001
202-727-1800

Jim Tait, Director
Governor's Energy Office
214 South Bronough Street
Tallahassee, Florida 32399-0001
904-488-6764

Paul Burks, Director
Office of Energy Resources
254 Washington Street, S.W.,
    Room 401
Atlanta, Georgia 30306
404-656-5176

Jerry M. Rivera, Director
Guam Energy Office
Office of the Governor
P.O. Box 2950
Agana, Guam 96910
671-734-4452

Maurice H. Kaya
Energy Program Administrator
Department of Business, Economic
    Development and Tourism
335 Merchant Street, Room 110
Honolulu, Hawaii 96813
808-587-3812

Robert W. Hopple, Administrator
Energy Resources Analysis Division
State House Mall
1301 North Orchard Street
Boise, Idaho 83720
208-327-7968

Mitch Beaver, Director
Division of Energy Conservation
    and Alternative Energy
Illinois Department of Energy and
Natural Resources
325 West Adams, Room 300
Springfield, Illinois 62704
217-785-2009

Amy Stewart, Director
Office of Energy Policy
Indiana Department of Commerce
One North Capitol, Suite 700
Indianapolis, Indiana 46204-2288
317-232-8939

Larry L. Bean, Administrator
Division of Energy and Geological
    Resources
Wallace State Office Building
Des Moines, Iowa 50319
515-281-8681/4308

Bill Ayesh, Supervisor
Energy Program
Kansas Corporation Commission
1500 SW Arrowhead Road
Topeka, Kansas 66604
913-271-3260

John Stapleton, Acting Director
Division of Energy
691 Teton Trail
Frankfort, Kentucky 40603-3249
502-564-7192

Diane Smith, Director
Energy Division
P.O. Box 44156
Baton Rouge, Louisiana 70804
504-342-2133

John Flumerfelt, Director
Energy, Policy and Planning
Maine State Planning Office
State House Station No. 38
Augusta, Maine 04333
207-624-6012

Gerald L. Thorpe, Director
Maryland Energy Administration
45 Calvert Street
Annapolis, Maryland 21401
301-974-3755

Paul W. Gromer, Commissioner
Division of Energy Resources
100 Cambridge Street, Room 1500
Boston, Massachusetts 02202
617-727-4732

Bob Forsythe, Acting Director
Michigan Public Service
    Commission
P.O. Box 30221
Lansing, Michigan 48909
517-334-6272

Kris Sanda, Commissioner
Public Service
American Center Building, 7th Floor
150 East Kellogg Boulevard
St. Paul, Minnesota 55101
612-296-7107

Andrew Jenkins, Director
Energy and Transportation Division
Mississippi Department of
    Economic and Community
    Development
510 George Street, Suite 101
Jackson, Mississippi 39202
601-359-6600

Bob Jackson, Director
Division of Energy
Department of Natural Resources
P.O. Box 176
Jefferson City, Missouri 65101
314-751-2254

Van C. Jamison, Administrator
Energy Division
1520 East 6th Avenue
Helena, Montana 59620-2301
406-444-6754

Bob Harris, Director
Nebraska State Energy Office
State Capitol Building, 9th Floor
P.O. Box 95085
Lincoln, Nebraska 68509-5085
402-471-2867

James P. Hawke, Director
Nevada Office of Community
    Services
Capitol Complex
400 West King, Suite 400
Carson City, Nevada 89710
702-687-5001/687-4990

Jonathan Osgood, Director
Governor's Office of Energy and
    Community Services
57 Regional Drive
Concord, New Hampshire
    03301-8506
603-271-2615

Scott A. Weiner, Commissioner
Energy Division
New Jersey Department of
Environmental Protection and
    Energy
401 East State Street, CN402
Trenton, New Jersey 08625
609-292-2885

John McGowan, Director
Energy Conservation and
    Management Division
2040 South Pacheco Street
Santa Fe, New Mexico 87505
505-827-5917

Charles R. Guinn, Deputy
    Commissioner
Policy Analyst Planning
New York State Energy Office
Two Rockefeller Plaza, 10th Floor
Albany, New York 12223
518-473-4377

Carson D. Culbreth, Director
North Carolina State Energy Office
430 North Salisbury Street
P.O. Box 25249
Raleigh, North Carolina 27611
919-733-2230

Shirley R. Dykshoorn, Director
Office of Intergovernmental Affairs
State Capitol, 14th Floor
Bismarck, North Dakota 58505
701-224-2094

Jocelyn F. DeLeon Guerrero,
Administrator
Commonwealth Energy Office
P.O. Box 340
Saipan, Mariana Islands 96950
670-322-9229/9236

Robert Garrick, Assistant Chief
Office of Energy Efficiency
Ohio Department of Development
77 High Street, 24th Floor
Columbus, Ohio 43215
614-466-6797

Sherwood Washington, Director
Division of Community Affairs and
    Development
Oklahoma Department of
    Commerce
P.O. Box 26980
Oklahoma City, Oklahoma 73126
405-841-9326

Christine A. Ervin, Director
Department of Energy
625 Marion Street, N.E.
Salem, Oregon 97310
503-378-4131

Brian Castelli, Executive Director
Pennsylvania Energy Office
116 Pine Street
Harrisburg, Pennsylvania 17101
717-783-9981

Carmen Vanessa Davila,
    Coordinator
Energy Program
Department of Consumer Affairs
P.O. Box 41059, Minillas Station
Santurce, Puerto Rico 00940
809-722-0960/721-0890

Scott Wolf, Director
Governor's Office of Housing,
Energy and Intergovernmental
    Relations
State House, Room 111
Providence, Rhode Island 02903
401-277-2850

Carl Roberts Jr., Director
Division of Energy, Agriculture and
    Natural Resources
1205 Pendleton Street, Suite 333
Columbia, South Carolina 29201
803-734-0445

Ron Reed, Commissioner
Governor's Office of Energy Policy
217 West Missouri Street, Suite 200
Pierre, South Dakota 57501-4516
605-773-3603

Cynthia Oliphant, Director
Energy Division
320 6th Avenue North, 6th Floor
Nashville, Tennessee 37219-5308
615-741-2994

Carol Tombari, Deputy Director
Energy Division
Governor's Office
P.O. Box 12428
Austin, Texas 78711
512-463-1931

Richard M. Anderson, Director
Utah Division of Energy
Three Triad Center, Suite 450
Salt Lake City, Utah 84110
801-538-5428

Scudder H. Parker, Director
Energy Efficiency Division
Vermont Department of Public
    Service
120 State Street
Montpelier, Vermont 05602
802-828-2811

Claudette Young-Hinds, Director
Virgin Islands Energy Office
Old Customs House
Frederiksted, Virgin Islands 00840
809-722-2616

Energy Director
Division of Energy
2201 West Broad Street
Richmond, Virginia 23220
804-367-1310

Richard H. Watson, Director
Washington State Energy Office
809 Legion Way, S.E.
Mail Stop FA 11
Olympia, Washington 98504-1211
206-956-2001

John F. Herholdt Jr., Manager
West Virginia Fuel and Energy
    Office
Building 6, Room 553
State Capitol
Charleston, West Virginia 25305
304-348-4010

Carol Cutshall, Director
Energy and Coastal Policy
101 South Webster, 6th Floor
P.O. Box 7868
Madison, Wisconsin 53707
608-266-8870

George Gault, Director
Division of Economic Development
Barrett Building, 4th Floor, North
    Wing
Cheyenne, Wyoming 82002
307-777-7284

Appendix 11
Highlights of the National Energy Policy Act of 1992

Compiled and written by Eric T. Sikkema, intern, Energy, Science, and Natural Resources Program, National Conference of State Legislatures. This summary is based on the conference report found in the *Congressional Record*, Oct. 5, 1992: 12056-12171. Title numbers may change when the law is codified.

The Energy Policy Act of 1992 (Public Law 102-486) focuses on the efficient production and consumption of energy in the United States. The legislation calls for increased use of renewable and domestic energy supplies, amends the Public Utility Regulatory Policies Act (PURPA), and reforms the Public Utility Holding Company Act (PUHCA). State efforts to bolster their economies and improve the quality of the environment through state legislation have served as a model for much of the legislation. The Energy Policy Act, as summarized in this appendix, presents new challenges and opportunities for state legislatures.

Title I - Energy Efficiency

Within two years of the law's enactment, states must incorporate specific energy-efficiency standards into their commercial building codes and consider upgrading the energy-efficiency codes for residential buildings. The Council of American Building Officials (CABO) and the American Society of Heating, Refrigerating, and Air-Conditioning Engineers (ASHRAE) are directed to develop voluntary energy codes for buildings.

The U.S. Department of Energy (DOE) is required to establish a voluntary home-energy rating system for comparing the relative energy efficiency of residential buildings. DOE also must develop and implement a pilot program in five states for energy-efficient mortgages (EEMS) to help new and current homeowners obtain mortgage financing for energy-efficiency improvements. The Department of Housing and Urban Development (HUD) is to establish thermal-insulation and energy-efficiency standards for manufactured housing. If the HUD standards are not in place within one year, then states may set their own standards.

Title I amends PURPA by adding federal requirements that (1) there be integrated resource planning (IRP) by utilities, (2) utility expenditures for demand-side management (DSM) be as profitable as investments in additional electricity generation, and (3) utility rates encourage investments for efficiency in the generation, transmission, and distribution of electricity. DOE is authorized to provide up to $250,000 to individual state regulatory bodies that have prepared plans for implementing demand-side management measures.

The law mandates the Western Area Power Administration (WAPA) to require that each of its customers who purchase power under long-term contracts implement integrated resource planning within three years (different requirements exist for smaller customers). If a customer does not meet the requirement or does not comply with an approved integrated resource plan, WAPA may place a 10 percent surcharge on sales to the customer in the first year and additional surcharges of up to 30 percent in subsequent years. WAPA may also reduce by 10 percent its allocation of power to non-complying customers.

The legislation provides support for a program to develop efficiency ratings and labeling for windows; establishes maximum flow rates for showerheads, faucets, and other plumbing products and efficiency standards for common fluorescent and incandescent lamps, as well as for commercial and industrial lighting and heating, ventilation, and air conditioning systems. It also requires DOE to establish "technologically feasible and economically justified" efficiency standards for small electric motors, utility distribution transformers, and high-intensity discharge lamps that would result in significant energy savings.

Title I authorizes cost-sharing grants to industry associations to support programs that improve energy efficiency in industry. The legislation also authorizes grants to encourage states, utilities, and industries to mutually assess various options for improving energy efficiency. Individual states may receive up to $1 million in financial assistance for energy-efficiency improvements to state and local government buildings. The federal funding will go to states committed to improving the energy efficiency of state and local government buildings, demonstrated by the adoption of energy efficient-codes for new buildings (at least as stringent as those established in the legislation) and a program for improving the efficiency of government buildings (financed by a revolving loan fund).

The title requires federal agencies to install by 2000 all energy and water conservation measures in their buildings that have a payback period of less than 10 years. To pay for the improvements, the law establishes funding options and offers incentives to agencies, including the use of energy performance contracting with the private sector.

Title II - Natural Gas

Title II requires the Federal Energy Regulatory Commission (FERC) to regulate natural gas imported from any country with which the United States has a free trade agreement in the same manner as it regulates domestic natural gas. It also includes a "sense-of-Congress" resolution that natural gas consumers and producers are best served by a competitive natural gas wellhead market.

Title III Alternative Fuels - General

Fleet requirements under Title III affect those who own or control at least 50 vehicles in the United States and fleets of at least 20 vehicles that are centrally fueled or are capable of being centrally fueled within a metropolitan area of 250,000 or more. Title III also requires the federal government to purchase at least 5,000 alternative fuel vehicles in 1993, 7,500 in 1994, and 10,000 in 1995. Beginning in 1996, 25 percent of vehicles acquired for federal fleets must be alternative fuel vehicles. That percentage is to increase annually to 75 percent in 1999.

Title IV - Alternative Fuels - Non-Federal Programs

Title IV exempts sales of natural gas for vehicles from regulation by FERC and by states (except for regulation to protect public safety). It also requires DOE to establish a public information program about the benefits and costs of using alternative fuels in motor vehicles. The Federal Trade Commission (FTC) is to issue rules on the labelling of alternative fuel vehicles and alternative fuels. The law authorizes the FERC to allow natural gas companies and electric utilities to recover expenses in advance for research and development of natural gas and electric vehicles through increased rates if it determines that program benefits exceed costs to existing and future ratepayers.

Title IV establishes programs and authorizes funding for financial assistance to states wishing to implement plans approved by DOE for accelerating the introduction of alternative fuels and alternative fuel vehicles, an alternative fuel program for mass transit, and a program to provide low-interest loans to purchase alternative fuel vehicles or to convert vehicles to alternative fuels.

Title V - Availability and Use of Replacement Fuels, Alternative Fuels, and Alternative Fueled Private Vehicles

Title V establishes requirements for alternative fuel vehicle purchases for those who produce, store, refine, process, store, or otherwise provide alternative fuels. Under this title, DOE is to establish a program to promote the development and use of domestic replacement fuels as a substitute for petroleum motor fuels.

Beginning in 1995, the following percentages of vehicles acquired for state fleets must be powered by alternative fuels: 10 percent in model year 1996, 15 percent in model year 1997, 25 percent in model year 1998, 50 percent in model year 1999, and 75 percent in model year 2000. States are permitted to develop plans (including conversion of existing vehicles) that meet or exceed the requirements. States that acquire more than the required number of alternative fuel vehicles in one year may receive credits that can be used to meet the requirements in the following year.

Title VI - Electric Motor Vehicles

The Electric Motor Vehicle Demonstration Program is established to provide $50 million over 10 years for joint ventures between public and private entities for the commercial demonstration of electric vehicles. The title also establishes an Electric Motor Vehicle Infrastructure and Support Systems Development Program for which $40 million over five years is authorized.

Title VII - Electricity

This title reforms PUHCA by establishing a class of "exempt wholesale generators" (EWGs)—corporate entities that may generate and sell electric power at wholesale without being subjected to the organizational restrictions mandated under PUHCA. A utility's power plants may be considered to be EWGs only if state regulators determine that doing so is in the public interest, will benefit customers, and does not violate state law. State regulators also are required to make similar determinations in order for an EWG to be able to sell electricity to a utility with which it is affiliated.

An EWG can apply to FERC for access to a utility's transmission lines. Access would be provided if FERC determines that it is in the public interest and if the price established for transmission services complies with criteria in the legislation. FERC is explicitly authorized to order transmission access (including enlargement of transmission capacity) for any utility, federal power marketing administration, or EWG, but it is prohibited from doing so if it would affect a utility's ability to provide reliable customer service.

Title VIII - High-Level Radioactive Waste

The U.S. Environmental Protection Agency (EPA) is directed to set radiation protection standards based on recommendations of the National Academy of Sciences for the maximum allowable doses of radiation to individual members of the public from high-level radioactive waste disposed specifically at the proposed Yucca Mountain, Nev., repository. The law requires the U.S. Nuclear Regulatory Commission (NRC) to modify its licensing criteria for the Yucca Mountain repository to be consistent with the new EPA standards and requires DOE to provide continuous oversight of the repository site in perpetuity to prevent activities that may release radiation. DOE also must conduct a study to determine whether current plans are adequate to manage additional high-level radioactive wastes that might be generated from nuclear power plants constructed in the future.

The title extends for two years the Office of the Nuclear Waste Negotiator, which is charged with finding a location for temporary storage (a monitored retrievable storage, or MRS, facility) of waste en route to the permanent repository.

Title IX - United States Enrichment Corporation

Title IX establishes the United States Enrichment Corporation to upgrade uranium for use in nuclear reactor fuel. The corporation replaces the existing DOE uranium enrichment program and is to operate on a profitable and efficient basis like any other business and to obtain, market, and sell enriched uranium. The corporation is directed to study options for transfer of ownership of the corporation to private investors and is authorized to purchase uranium from the former Soviet Union.

Title X- Remedial Action and Uranium Revitalization

Regarding the cleanup of uranium and thorium processing sites, DOE is required to reimburse owners (licensees) of the facilities in a timely manner, provided the request for reimbursement is supported by reasonable documentation.

Title X establishes the National Strategic Uranium Reserve to purchase domestic uranium and to efficiently produce enriched uranium in a manner consistent with "sound business practices." It also reauthorizes the Uranium Mill Tailings Radiation Control Act for two years to continue to clean up radioactive mill tailings.

Title XI - Uranium Enrichment Health, Safety, and Environment Issues

The law specifies the Nuclear Regulatory Commission's role in regulating uranium enrichment facilities and establishes a Uranium Enrichment Decontamination and Decommissioning Fund for the eventual cleanup of existing DOE enrichment facilities.

Title XII - Renewable Energy

Title XII authorizes incentive payments to utilities for renewable energy production. For a period of 10 years after the law's enactment, a utility is eligible to receive payments of 1.5 cents per kilowatt hour for 10 years for energy it supplies from new or purchased facilities that generate electricity derived from solar, wind, biomass, or geothermal energy resources.

It authorizes $50 million for fiscal year 1994 for a demonstration and commercial application program for renewable-energy and energy-efficiency technologies and provides for an interest rate "buy-down" program to reduce the financial risk to private lenders who are asked to fund renewable projects that would otherwise be classified as too risky.

Title XII also establishesan interagency working group to inform foreign countries of the benefits of renewable-energy and energy-efficiency policies and creates a training program to instruct representatives from developing countries about the operation of U.S. renewable-energy and energy-efficiency products and technologies. Other provisions under this title set goals for the national Alcohol From Biomass Program, mandate a National Renewable and Energy Efficiency Management Plan, authorize an awards program to recognize advancements in renewable energy applications, and require DOE to conduct a study of the relative tax treatment of renewables compared with conventional power plants.

Title XIII - Coal

Title XIII authorizes numerous research, development, and demonstration projects (including joint ventures with private industry) for commercial applications of coal-related technologies. It extends the DOE's Clean Coal Technology Program and establishes an interagency task force to promote and expand the transfer of clean coal technologies to developing countries.

The Interior and Energy departments are instructed to identify states that have (1) ownership disputes that impede methane gas development, (2) no regulatory procedures for coalbed methane gas development, and (3) no extensive development of coalbed methane gas. States designated as "affected states" are given three years to develop programs for resolving ownership disputes between mineral rights holders and land owners. If an affected state does not comply with the requirements, the federal government could impose "forced pooling" arrangements, whereby it would take control of the program and allow drilling for methane gas to continue with proceeds placed in escrow until legal determination of ownership is resolved. Seven states designated in the legislation as affected states are Illinois, Indiana, Kentucky, Ohio, Pennsylvania, Tennessee, and West Virginia. Excluded from future designation are Alabama, Colorado, Louisiana, Montana, Mississippi, New Mexico, Utah, Virginia, Washington, and Wyoming.

The title prohibits the Department of the Interior from implementing a proposed rule under which the federal government would have to allow coal companies to conduct mining operations in areas such as national parks and forests or otherwise compensate them for the value of the coal that they could have mined.

Title XIV - Strategic Petroleum Reserve

The president is permitted to undertake measures to increase the size of the strategic petroleum reserve (SPR) to one billion barrels as rapidly as possible and to authorize the sale of oil from the SPR in response to severe oil price increases that would adversely affect the national economy.

Title XV - Octane Display and Disclosure

Title XV redefines automotive fuel to include all liquid fuels, preempts state law that conflicts with federal law but does not allow state investigative or enforcement actions, requires an EPA study of the anti-knock characteristics of fuels, and requires an FTC study of uniform national labels on fuel pumps.

Title XVI - Global Climate Change

Title XVI directs DOE to conduct a feasibility study of the economic, energy, and environmental implications of stabilizing U.S. greenhouse gas emissions by 2005, the progress of the United States compared with other countries in reducing emissions of greenhouse gases, and the implications of reducing 1988 carbon dioxide emissions by 20 percent by 2005. It requires that future national energy strategies include a least-cost energy plan prepared by DOE to address considerations for economic, energy, social, and environmental costs and benefits. It also establishes a program for transferring technology to other countries and creates a Global Climate Change Response Fund for U.S. assistance to global efforts to lessen and adapt to global climate change.

Title XVII - Additional Federal Power Act Provisions

Title XVII prohibits hydroelectric power developers from using the right of eminent domain to acquire rights-of-way for federally licensed projects that cut across state or local government lands set aside as parks, recreation areas, or wildlife refuges prior to the law's enactment. For lands set aside after enactment, hearings must be conducted, and FERC must make a determination that the project will not interfere with the purposes of that land before developers may declare eminent domain.

Title XVIII - Oil Pipeline Regulatory Reform

Within 18 months after enactment, FERC is required to promulgate a rule to streamline procedures to avoid unnecessary regulatory costs and delays and is further directed to identify other procedural changes that could be altered to expedite the process.

Title XX - General Provisions: Reduction of Oil Vulnerability

This title requires a number of programs to strengthen national energy security, including programs for enhanced oil recovery, increased oil shale extraction, and increased natural gas supply. In the transportation sector, the programs are aimed at improving alternative and natural gas fuels technologies, developing technology for increased automotive fuel economies, and a study of the costs and benefits of telecommuting.

Title XXI - Energy and Environment

Title XXI authorizes funding for programs directed at increasing energy efficiency and the use of renewable energy and providing cost-effective options for the generation of electricity from renewable energy sources. It also authorizes money for an advanced light-water reactor program and a prototype demonstration of advanced reactor technology.

Title XXII - Energy and Economic Growth

Title XXII establishes a variety of programs, including a national advanced materials initiative, a vigorous effort to promote research into energy development, and a math and science education program directed at low-income and first-generation college students.

Title XXIII - Policy and Administrative Positions

This title requires DOE to submit to Congress for review a management plan for all projects involving $100 million or more; establishes an Energy Research, Development, Demonstration, and Commercial Application Advisory Board; requires DOE to prepare a management plan for the goals identified under Title XX; provides guidelines to DOE for selecting contractors for decommissioning, storage, and disposal of nuclear waste; and establishes eligibility requirements for companies to receive financial assistance under titles XX through XXIII.

Title XXIV - Non-Federal Power Act Hydropower Provisions

Under Title XXIV, FERC is prohibited from issuing any license for the construction of new hydroelectric power projects located within the National Park System if the project "would have a direct adverse effect" on those lands. It permits the use of FERC-certified, third-party contractors to prepare impact statements required as part of the FERC hydropower licensing process.

The Interior and Energy departments are directed to identify opportunities to increase power production and efficiency at federal hydroelectric projects, as well as to study the potential for adopting water conservation measures while not endangering water supplies for fish and wildlife.

Title XXV - Coal, Oil, and Gas

This title directs the U.S. Geological Survey to establish hot dry rock geothermal programs; establishes conditions for coal remining under the Surface Mining Control and Reclamation Act of 1977 (SMCRA); amends SMCRA to require mining companies to compensate (1) owners of residential and commercial buildings damaged by subsidence and (2) homeowners whose drinking water supplies are adversely affected by underground mining activities; requires that mineral leasing revenues from acquired lands be shared with states; and makes grants to the Navajo, Hopi, Northern Cheyenne, and Crow tribes to help them develop surface coal mining and reclamation programs.

Title XXVI - Indian Energy Resources

Title XXVI requires the Interior and Energy departments to establish a demonstration program to assist Indian tribes to pursue energy self-sufficiency and increase development of energy resources located on Indian reservations. The Department of the Interior is to make grants to Indian tribes to assist them in the development, implementation, and enforcement of tribal energy development laws and regulations. The title also establishes an Indian Energy Resources Commission and authorizes financial assistance to Indian governments for energy-efficiency and renewable-energy projects.

Title XXVII - Insular Areas Energy Security

Title XXVII authorizes financial assistance to insular area governments for projects to study and encourage energy-efficiency and renewable-energy measures as a means to increase their energy security.

Title XXVIII - Nuclear Power Plant Licensing

The law gives the NRC authority to issue combined licenses to build and operate new nuclear power plants. The process is streamlined by not requiring new facilities to receive a second permit to operate the facility once it has been constructed. A formal public hearing process is required before the combined license can be issued, but a public hearing would no longer be required after construction unless there is convincing evidence that the plant is not meeting the license requirements. If a post-construction hearing is held, the NRC may allow the plant to operate on an interim basis if it determines that public health and safety are not endangered. Another section of the bill requires the DOE to promote the development and deployment of advanced technologies for nuclear reactors.

Title XXIX - Additional Nuclear Energy Provisions

This title revokes a previous NRC policy statement that permitted designation of some low-level radioactive waste as "below regulatory concern," thereby deregulating its disposal. Had the policy been implemented, certain low-level radioactive waste could have been incinerated or disposed of in landfills. States are authorized to regulate the disposal of any low-level radioactive waste that may be deregulated by the NRC in the future.

Other provisions under this title expand protection for "whistleblowers" (employees who report safety problems at nuclear facilities) and require a study by the president and the NRC on the safety of shipments of plutonium by sea.

Title XXX - Miscellaneous

The final title (1) addresses guidelines and procedures for research and development agreements between industry and federal laboratories, (2) abolishes the Office of Federal Inspector of Construction for the Alaska Natural Gas Transportation System, (3) encourages the use of geothermal heat pumps, (4) requires a study and pilot program on the use of energy futures for fuel purchases to the strategic petroleum reserve, (5) requires the National Academy of Sciences to conduct a study of energy subsidies, (6) requires a tar sands study, and (7) establishes a Consultative Commission on Western Hemisphere Energy and Environment.